BUILDING AMERICA
THEN AND NOW

THE TELEPHONE
WIRING AMERICA

BUILDING AMERICA: THEN AND NOW

BUILDING AMERICA
THEN AND NOW

THE TELEPHONE
WIRING AMERICA

JOHN MURPHY

CHELSEA HOUSE
PUBLISHERS
An imprint of Infobase Publishing

The Telephone: Wiring America

Copyright © 2009 by Infobase Publishing

Chelsea House
An imprint of Infobase Publishing
132 West 31st Street
New York, NY 10001

Library of Congress Cataloging-in-Publication Data
Murphy, John, 1968–
   The telephone : wiring America / by John Murphy.
     p. cm. — (Building America : then and now)
   Includes bibliographical references and index.
   ISBN 978-1-60413-068-3 (hardcover)
   1. Telephone—United States. 2. Telecommunication—United States. I. Title.
   HE8815.M87 2009
   384.60973—dc22   2008025546

Text design by Annie O'Donnell
Cover design by Ben Peterson

Printed in the United States of America

Bang NMSG 10 9 8 7 6 5 4 3 2 1

This book is printed on acid-free paper.

CONTENTS

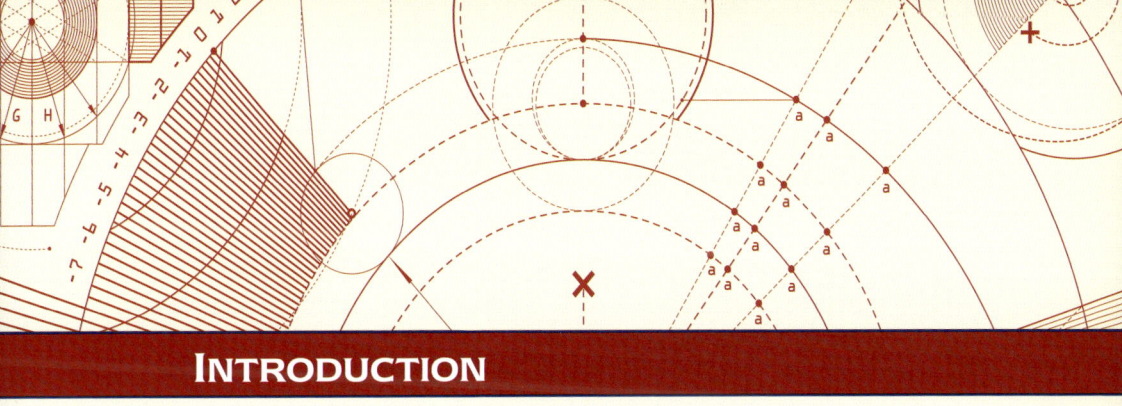

A Worldwide Web

Lifeline of the lonely and lifeblood of the busy, [it] is taken for granted, and for good reason. . . . By bringing about a quantum leap in the speed and ease with which information moves from place to place, it has greatly accelerated the rate of scientific and technological change and growth in industry. Beyond doubt it has crippled if not killed the ancient art of letter writing. It has made living alone possible for persons with normal social impulses. . . . Certainly it has extended the scope of human conflicts, since it impartially disseminates the useful knowledge of scientists and the babble of bores, the affection of the affectionate and the malice of the malicious. . . . [It] is our nerve-end to society.

The preceding quotation sounds as if it were written just a moment ago by a technology, media, or social critic at the dawn of the World Wide Web, at once afraid and in awe of Internet technology. In fact, these are not the words of a hand-wringing blogger. They were written more than 30 years ago by John Brooks, about a much earlier technological phenomenon that profoundly altered and reconceived our world every bit as much as the Internet has. Brooks was writing of an invention now viewed almost as a quaint relic of the analog age, yet one that revolutionized science, society, business, and technology. This invention made possible many of the modern world's most stunning achievements, including radio, motion pictures, high-fidelity recording, television, cellular technology, and computer and Internet technology.

Indeed, this seemingly humble but astoundingly powerful and transformative invention—and the far-reaching network it spawned—is still a crucial part of our daily lives, used by billions of people worldwide. It is the telephone, and it is impossible to imagine the technological and social achievements of the twentieth century without it, much less the innovations and capabilities we have taken with us into the new millennium.

Long before there was the Interstate Highway System or the World Wide Web to connect the United States, the nation and its people were united by the common bond of telephone wires crisscrossing the country. Though we now take the humble landline telephone for granted, its invention and subsequent development by telephone industry engineers made possible some of our most cherished tools and toys of today: radio, movies, television, satellites, microwaves, radar, computers, the Internet, digital transmission, MP3 players, and the great successor to Alexander Graham Bell's telephone—the cell phone.

In what it made obsolete, what it made possible, and how much of the world and its culture it transformed, the telephone ranks as one of the most important inventions in the history of human civilization. The telephone remade the world in an entirely new image, and our world today would be unrecognizable to us had the telephone not been invented. It is fitting that it was first introduced to Americans in 1876, on the one hundredth anniversary of the nation's birth. This invention did more to unite and liberate widely scattered, divided, and isolated Americans than any previous document, law, treaty, roadway, railway, invention, or innovation could. People were now connected to one another by electrical wires that magically carried the tones of

The telephone ushered in a new era of communication, commerce, and social interaction. Without having to leave home or write a lengthy letter, people all over the United States could connect with faraway friends and family members. Referred to as one of the greatest inventions in history, the telephone soon became a fixture in the American home.

the human voice. Ordinary Americans could talk to one another whenever they got the urge. Thanks to Alexander Graham Bell and any number of anonymous engineers throughout the years, all someone had to do was pick up the telephone.

CHAPTER 1

The Dawn of Telecommunications

Electricity began to revolutionize human communications beginning in the 1820s and 1830s. The first telegraph machine was patented in 1837 by the Englishmen William Cooke and Charles Wheatstone. The device was improved upon by the American inventor Samuel Morse (for whom Morse code is named). The telegraph represented a huge leap forward, allowing—for the first time in human history—the direct, instantaneous, and perfectly comprehensible communication between two people across vast distances.

Previous to this startling breakthrough, human long-distance communication was slow, unreliable, and fraught with misunderstandings and missed messages. At the time of the invention of the telegraph, the early nineteenth century's most advanced long-distance communications systems were handwritten messages carried via the Pony Express, steam and sailing ships, and canal and river boats. Short railroad lines were just beginning to appear in the United States and England, but they were still decades away from evolving into the national systems of "steel roads" that they would become by the end of the nineteenth century, which allowed for more rapid cross-country movement of goods and information.

INSTANT COMMUNICATION

From the perspective of our modern world, steamships, horse-mounted postal carriers, and even locomotives provided only marginal improvements in the speed and quality of communications over older forms of message conveyance, which included smoke signals, running messengers, couriers, stagecoaches, semaphore signals (an alphabetic code created by the positioning of mechanical arms or waving flags), and even carrier pigeons. The telegraph, harnessing the recently discovered power of electromagnetism, helped usher in the age of electricity, which would also see the invention of the lightbulb, motors, engines, and dynamos. Perhaps most importantly, it would also introduce the world's people—long accustomed to waiting for days, weeks, months, or even years for news and information of both a public and private nature—to the thrilling pleasure of immediate communication.

Hundreds of small, local telegraph companies popped up nationwide, spurring frenzied regional competition and a crazy-quilt telegraph network. Ten years after the completion of Morse's telegraph line from Baltimore to Washington, D.C., 23,000 miles of telegraph line had been strung, exceeding the total mileage of the nation's railroad tracks. Almost all of the eastern United States enjoyed telegraph service, and some cities had their pick of

Opposite, an advertisement for the Pony Express. Before the telephone and the transcontinental railway, correspondence was sent via telegraph, ship, or courier. One of the most popular and well-known courier services, the Pony Express employed men on horseback to deliver the nation's mail. Even with their fastest riders, however, the Pony Express still took about 10 days to deliver a letter across the country.

PONY EXPRESS !

CHANGE OF

TIME !

REDUCED

RATES !

10 Days to San Francisco!

LETTERS

WILL BE RECEIVED AT THE

OFFICE, 84 BROADWAY,

NEW YORK,

Up to **4** P. M. every TUESDAY,

AND

Up to **2½** P. M. every SATURDAY,

Which will be forwarded to connect with the PONY EXPRESS leaving
ST. JOSEPH, Missouri,

Every WEDNESDAY and SATURDAY at 11 P. M.

TELEGRAMS

Sent to Fort Kearney on the mornings of MONDAY and FRIDAY, will connect with **PONY** leaving St. Joseph, WEDNESDAYS and SATURDAYS.

EXPRESS CHARGES.

LETTERS weighing half ounce or under...............$1 00
For every additional half ounce or fraction of an ounce 1 00
In all cases to be enclosed in 10 cent Government Stamped Envelopes,

And all **Express CHARGES** Pre-paid.

☞ PONY EXPRESS ENVELOPES For Sale at our Office.

WELLS, FARGO & CO., Ag'ts.

New York, July 1, 1861.

SLOTE & JANES, STATIONERS AND PRINTERS, 95 FULTON STREET, NEW YORK a dozen service providers. Indeed, there were so many telegraph service providers that communications actually became delayed. Any given message might have to be routed through the wires of a dozen or more telegraph companies to reach its destination.

WESTERN UNION

Consolidation of the telegraph industry began once some investors realized just how much money could be made by operating telegraph offices. Consolidation also helped impose greater clarity and efficiency on the chaotic national "system" of telegraph wires and service providers that was so disorganized it could scarcely be referred to as a true system. By the 1860s, two large companies in particular—the American Telegraph Company and the Western Union Telegraph Company—had come to control the industry in the United States. They bought up the smaller regional and local companies and, between them, owned thousands of telegraph offices nationwide and hundreds of thousands of miles of telegraph lines.

Soon, Western Union ran telegraph lines from Missouri to California, and the western and eastern halves of the country were finally linked by a single, dedicated, rapid line of communication. This development represented the death knell for one of the nation's celebrated, preindustrial forms of "rapid" communication—the Pony Express. Not only were personal and business communications being sent and delivered with unimaginable rapidity, but so too was news and information. Indeed, the Associated Press (AP) became the nation's first telegraphic news company. The AP still exists today and, along with several other similar news organizations, is still known as a "wire service," which provides breaking news that is said to "come across the wires."

Seemingly, there was no stopping the march of telegraphic progress. In 1866, following an aborted effort in 1858, the first lasting transatlantic telegraph cable—which contained telegraph wires wrapped in waterproof insulating rubber—was laid across

Samuel F. B. Morse *(above)* created a simpler version of the telegraph and an accompanying system of communication known as Morse code. Using a series of dots and dashes, Morse code allowed telegraph operators to transmit messages more efficiently. As demand grew, transmission cables were soon strung up alongside the tracks of the transcontinental railroad for instantaneous coast-to-coast messaging.

the ocean, allowing for instantaneous communication between the United States and England, and beyond to the European continent. This venture was aided by the technical contributions of the English scientist Lord Kelvin, who would one day witness with wonder the first major public demonstration of Alexander Graham Bell's telephone.

This is the rapidly changing, swiftly expanding nation in which Alexander Graham Bell arrived from Scotland (after a brief stay in Canada) in 1871. The United States and the Western world were growing exponentially, thanks to industrialization, yet they also were growing smaller and more tightly connected thanks to the modern communication era ushered in by Morse and others. It was now time for Bell and his peers to take the telegraph to the next level and introduce the world to innovations that would transform society in even more profound ways than Morse's invention had.

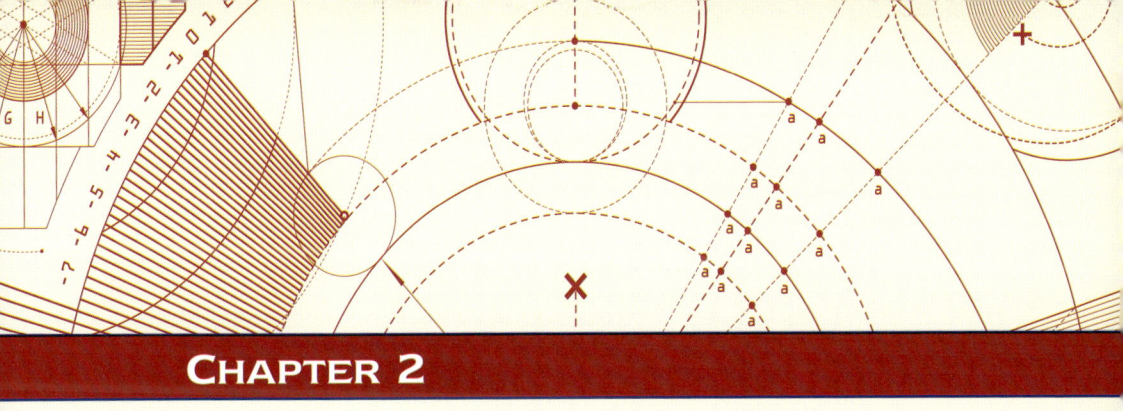

CHAPTER 2

Bell's Harmonic Telegraph

Alexander Graham Bell was born in Edinburgh, Scotland, on March 3, 1847, three years after Morse completed his ground-breaking Baltimore-to-Washington telegraph line. The family he was born into was steeped in the art and science of speech and sound creation. Bell's grandfather, Alexander Bell, was a shoemaker turned actor who eventually opened an elocution school to help cure students of stuttering. Elocution is the study and practice of speech formation and transmission. It involves observing and perfecting how a speaker forms sounds and words and then vocalizes them correctly.

THE INVENTIVE BELLS

Melville Bell, Alexander Bell's son, followed in his father's foot-steps and became a professor of elocution at the University of London, where he specialized in teaching deaf students how to speak clearly and intelligibly. Melville invented a code of "Visible Speech," which featured written symbols that indicated exactly

[ENGLISH ALPHABET OF VISIBLE SPEECH,

Expressed in the Names of Numbers and Objects.]

[Pronounce the Nos.] [Names.]	[Name the Objects.]		[Name the Objects.]	
1.				
2.				
3.				
4.				
5.				
6.				
7.				
8.				

[EXERCISE.]

One by one.
Two or three.
Four at once.
Five o'clock.
Half-past six.
Seven-thirty.
Eight to nine.
Ten or twelve.
Twice two, four.
Twice three, six.
Four and four, eight.
Nine and two, eleven.
Twice or thrice.

Two, a couple.
Twelve, a dozen.
Twenty, a score.
A book-case.
A few books.
New book-shelves.
A silver watch.
A gold watch.
The watch-key.
A good saw.
Cap and feather.
Tongs and shovel.
Sugar-tongs.

A hunting whip.
A table lamp.
A bunch of onions.
Corns and bunions.
A ship's boat.
A sailing boat.
Cart and horse.
A round tent.
Rows of houses.
A dog-kennel.
A little monkey.
A pretty cage.
A green canary.

Melville Bell, father of Alexander Graham Bell, developed a method of communication for the deaf called Visible Speech. This system was based on a series of symbols representing vocal sounds that made up speech, allowing those who could not hear to interact with mainstream society. *Above*, the English alphabet of Melville Bell's Visible Speech.

what positions a speaker's throat, tongue, and lips should be in and how they should move to vocalize a specific sound correctly. Visible Speech allowed deaf students to form words clearly even if they couldn't hear the sounds they were making. Melville Bell believed that other forms of communication available to deaf students, such as sign language, only isolated them further from the world of the hearing.

Alexander Graham Bell inherited his father's inventiveness and interest in sound and speech. Challenged by his father to invent a machine that could talk, Alexander and his older brother fashioned a model of a human skull out of rubber and placed within in it a voice box made of metal, rubber, and an actual lamb's larynx. When the brothers blew through this artificial vocal cord system using a bellows taken from a parlor organ, the noise that emitted from the skull's mouth sounded something like the word *mama*.

A BRILLIANT MISTAKE

At the age of 16, Alexander Graham Bell left home to serve as a music and elocution tutor at a boys' school in Elgin, England. In between his teaching duties, he began to conduct speech and sound experiments that focused on proper vowel-sound creation and resonance and the necessary positioning of the mouth and tongue. When Melville Bell heard about his son's experiments, he told Alexander about similar research conducted by a German scientist named Hermann von Helmholtz. In his book *On the Sensations of Tone*, Helmholtz described his success in producing vowel sounds through the use of electrified tuning forks. Alexander Graham Bell, who was intrigued by this idea, found a copy of the book. The book was written in German, however—a language he did not know. It also contained confusing illustrations and concerned itself mainly with electricity, a field of study that was foreign to Bell at this point.

As a result, Bell had great difficulty reading the book and understanding the nature and processes of Helmholtz's work. In fact, Bell's inability to decipher the foreign text led to a large—and profoundly important—misunderstanding. Bell thought that Helmholtz was reporting that he had transmitted these artificially generated vowel sounds over telegraph wires, which was not at all the case. Bell's lack of understanding, however, planted the seed of an idea that would result in an entirely new field of knowledge. Bell suddenly believed that it was possible to transmit tones and sounds through electric wire and have it be heard clearly on the other end.

Bell would not realize his error for several years, but by then he had already tried and failed to copy Helmholtz's supposed experiment and had begun to test his own theories on the electrical and acoustical technology that would result in the telephone. Bell himself recognized just how important and useful his misunderstanding was in determining the course of his future work, as quoted by H.M. Boettinger in *The Telephone Book: Bell, Watson, Vail, and American Life, 1876–1976*: "I thought Helmholtz had done it and that my failure was due only to my ignorance of electricity. It was a very valuable blunder. It gave me confidence. If I had been able to read German, I might never have begun my experiments in electricity."

In the long history of human invention, believing that something is possible often turns out to be at least as important, if not more so, than knowing exactly how to make it possible. Believing that something is possible is also an essential first step to making it a reality.

MUSICAL MESSAGES

In 1870, following the death of two of Bell's brothers, both of whom had contracted tuberculosis, the Bell family moved to Brantford, Ontario, in Canada. Alexander soon relocated to Boston, where he taught his father's Visible Speech methods at several schools for the deaf in Massachusetts and Connecticut. Within two years he had become a professor of vocal physiology at Boston University, was preparing to open his own speech and elocution school, and was giving lectures throughout the eastern half of the United States on deaf education and speech mechanics.

Yet even during this time of intense busyness, Bell continued to experiment with the ideas sparked by Helmholtz's book during his free time at night. He had begun work on a device he called a "harmonic telegraph" that would allow multiple messages to be sent at the same time and received over a single wire. He discovered that, if he sang the G note very close to the strings of a piano, the G string would vibrate in response as if gently struck. Bell concluded, as quoted by Herbert N. Casson in *The History of the Telephone*, that "we may some day have a musical telegraph, which will send as many messages simultaneously over one wire as there are notes on [the] piano."

Unlike the conventional telegraph, the wires would not simply convey electric pulses that caused the receiver to click out the dots and dashes of a message. Bell's telegraph would carry musical tones. The transmitter would have several differently tuned tuning forks, each of which emitted a distinct tone. The telegraph key would switch one of the tuning forks on and off, transmitting a coded series of musical notes. The telegraph receiver would have the same array of tuning forks, and each receiver tuning fork would vibrate to the notes sent by its mate at the transmitter. The receiving tuning fork would emit those notes just as they had been sent. Each receiving tuning fork could pick up only the messages sent by its mate. It could not pick up those messages transmitted by a differently tuned fork.

In this way, several transmitting tuning forks could send messages simultaneously along the same wire to a receiver, and the multiple messages would not get crossed, confused, or garbled. This would represent a significant development over the single-message telegraph. The nation, especially its cities, was becoming tangled with millions of poles of web-like telegraph wires. If the technology could be developed for a single wire to carry multiple messages simultaneously, the infrastructure could be simplified and made more efficient, and the unsightly and at times dangerous wires could be reduced in number.

The tuning forks were eventually replaced by thin steel or iron strips, called *reeds*, that operated on the same principle. Along the way, Bell rejected an alternate idea for the harmonic telegraph that would have involved someone uttering words into a speaking trumpet at the transmitting end. The vibrations would generate an electrical current, which in turn would travel over an electrical wire. At the receiving end, a harp would vibrate to this current and reproduce the speaker's message in musical tones.

GAINING PATRONS

In addition to his schoolteaching, lectures, and private experiments, Bell also tutored; he offered private, one-on-one instruction to deaf children of mostly wealthy parents. Through his tutoring service he met two men who would become the most important and influential financial backers of his experimental work. The first of these was Thomas Sanders, a prominent Massachusetts leather merchant, whose son, George, had been born deaf.

The second important new figure in Bell's life was Gardiner Greene Hubbard, a prominent, wealthy Boston lawyer. In addition to his wealth and social connections, Hubbard had a keen interest in cutting-edge science and a willingness to finance new ideas. He was instrumental in bringing gas streetlights and a clean and reliable water supply to Cambridge, Massachusetts. He also helped provide a streetcar system for Boston. Hubbard served as the very first president of the National Geographic Society and as a board member of the Smithsonian Institution, and he was an investor in railroads, gas-lighting schemes, and other "high-tech" innovations of the day. His daughter Mabel became deaf at the age of four or five, following a bout of scarlet fever. She was a teenager when Bell began tutoring her.

Hubbard shared Bell's feelings about reducing the isolation of deaf people by teaching them how to vocalize properly, rather

Alexander Graham Bell *(above with his family)* fell in love with one of his students, Mabel Hubbard. Mabel's father, Boston lawyer Gardiner Greene Hubbard, became Bell's patron and business partner.

than adopting alternative—and to their minds unintentionally even more isolating—forms of communication such as sign language. As the president of the Clarke School for the Deaf in Northampton, Massachusetts—one of the schools in which Bell taught—Hubbard was steeped in the same issues of deafness, hearing, acoustics, and speech that were Bell's focus and were central to his ongoing harmonic telegraph experiments. Impressed by Bell's theories, teaching, and success with Mabel's instruction, Hubbard began to consider Bell's experiments as his next great investment.

Indeed, Bell, Hubbard, and Sanders began to speak seriously about creating a partnership that would result in the invention, patenting, manufacturing, and selling of Bell's harmonic telegraph. In the fall of 1874, Hubbard even combed through patent records to make sure that no one currently held any patent for a similar idea. Once he was assured that no such patent yet existed, he and Sanders agreed to fund Bell's work on the understanding that they would each hold an equal share in any patents obtained and the income generated by them.

All three men were motivated as much by altruism, idealism, and personal and emotional needs as by financial and commercial considerations. They wanted to become rich and maybe even famous, certainly, but they also all shared a personal connection to the hearing impaired and had a unique insight into what was then a lonely and isolating world for the deaf. Hubbard and Sanders had beloved deaf children. Bell had grown up in a family dedicated to education and socialization of the deaf, and his mother had become deaf during his childhood. In addition, Bell was beginning to fall in love with Mabel Hubbard, who would become his wife in the momentous year of 1876. The enthusiastic and passionate financial backing of Hubbard and Sanders meshed well with Bell's own energy, industry, and idealism. With their help, encouragement, and occasional bullying, Bell began a new round of experiments that would set him firmly on the road to the invention of the telephone.

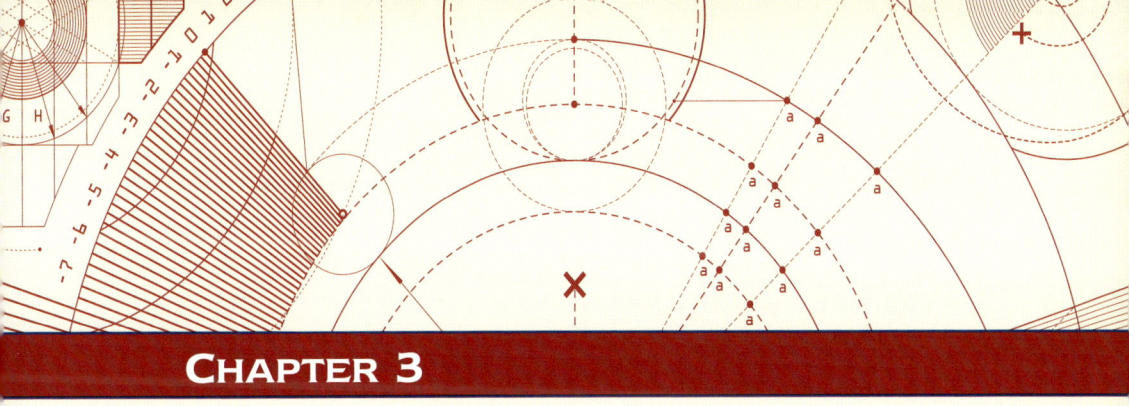

CHAPTER 3

A Neck-and-Neck Race

By carefully studying the human ear, Bell learned that a tiny membrane could move bones far heavier than itself, resulting in the accurate reproduction of sounds that were both very loud and extremely soft. He then concluded that a single membrane could similarly capture the vibrations of sound at the transmitting end of the harmonic telegraph and accurately reproduce these sounds of varying pitches, tones, and volume at the receiving end.

Bell began to wonder if his harmonic telegraph could also use a single membrane to create the necessary electrical pulses rather than his complex system of variously tuned tuning forks. These membrane-conveyed electrical pulses would copy the sound waves generated on the transmitting end, which could then be translated into actual sound on the receiving end. The next logical question Bell asked himself was, if all sound can be reproduced faithfully and accurately in this manner, why couldn't the human voice also be transmitted and received over electrical wires? The conceptual leap had been made, and Bell was now intent on creating something far more wondrous and useful than the harmonic telegraph: the telephone.

THOMAS WATSON

To translate his conceptual leap into a practical breakthrough, Bell would require not only a funding stream provided by his financial backers but also the nuts-and-bolts mechanical and electrical engineering expertise that he did not possess. Bell needed a knowledgeable and trusty assistant who could apply his mechanical and electrical know-how to Bell's lofty ideas, making them workable and practical in the real world. Bell would soon find this man.

In 1875, Bell began to perform his experiments at a Boston electrical shop run by Charles Williams Jr. The shop manufactured electrical apparatus like telegraphs, burglar alarms, and early forms of electrical doorbells. Williams employed a young, skilled machinist named Thomas A. Watson. Bell and Watson first met when Bell stormed into the shop with a defective harmonic telegraph device that had not been built according to his specifications. Watson quickly made the necessary repairs, following an explanation by Bell of what the device was and what he hoped it could do. Watson was captivated by Bell's ideas, especially the "autograph telegraph," which would use harmonic telegraph technology to transmit handwriting and pictures (an early fax machine of sorts). Soon, Watson became Bell's full-time assistant, and the two began to spend long hours together, trying—and mostly failing—to first perfect a working harmonic telegraph and, ultimately, the telephone.

A RIVAL EMERGES AND PRESSURE INCREASES

Bell felt an increasing sense of urgency to create a reliable working prototype of the harmonic telegraph. First, money was tight, and he knew he could not rely on the patience or funds of Hubbard and Sanders forever. In order to focus more intently on his

Restless and easily bored, Thomas A. Watson *(above)* tried several occupations before finding work at a local machine shop. His proficiency with electrical machines got him hired as the assistant of Alexander Graham Bell. Watson worked with Bell for many years and later established successful careers in shipbuilding, geology, and acting.

experiments and achieve breakthrough results, he cut down on his lecturing, teaching, and tutoring. Due to this loss of income, money became even tighter, and he was occasionally forced to borrow cash from his assistant, Watson. His anxiety was made far more acute, however, by the knowledge of competition.

Bell had come to learn that another inventor—Elisha Gray, cofounder of the Western Electric Company—was working in Chicago on an almost identical device. There was every reason to believe that Gray, who was generally considered the leading electrical engineer in the United States, would invent a working model of the harmonic telegraph before the far less experienced and knowledgeable Bell. If Gray completed a working model of a harmonic telegraph and received a patent for it first, not only would he receive all of the credit and acclaim for the invention, he also would receive all of the profits generated by it for the duration of the patent's protection. In late 1874, Bell wrote to Hubbard and Sanders (as quoted by Seth Shulman in *The Telephone Gambit*), "It is a neck-and-neck race between Mr. Gray and myself who shall complete an apparatus first. He has the advantage over me in being a practical electrician—but I have reason to believe that I am better acquainted with the phenomena of sound than he is—so that I have an advantage there . . . I feel I should be seriously ill should I fail in this now I am so thoroughly wrought up."

Bell realized that he needed to step up his efforts and rely on as much expert advice as he could get. He also hoped to generate excitement and additional funding for his invention; he shopped it around (unsuccessfully) to Western Union and, in March 1875, demonstrated it for the secretary of the Smithsonian Institution, Joseph Henry, one of the world's leading innovators in the field of electromagnetism. Henry's work with insulated electrical wires and the boosting of electromagnetic power led directly to the invention of the telegraph. Henry had provided some well-timed advice to Morse when he had gotten stuck while inventing his telegraph, and he would offer the same favor to Bell. He was unimpressed by Bell's harmonic telegraph but was intrigued by his passing references to the possibility of human speech transmitted via electrical wires. He encouraged Bell to scrap the harmonic telegraph immediately and instead redirect all of his energies and ingenuity to the potentially far more innovative invention of a human voice transmitter.

Flattered and gratified by the enthusiasm and encouragement of such a well-known scientist as Joseph Henry, Bell was keen to explore more fully the potential for voice transmission, rather than wait until the successful completion of a harmonic telegraph. In an uncharacteristic burst of brash self-confidence, Bell declared, according to Casson, "If I can make a deaf-mute talk, I can make iron talk." This would also allow him to separate himself from Elisha Gray and worry less about competition and the ticking clock, because they would be working on different devices and technologies. Yet, given his current underemployment and lack of income, Bell was more constrained than ever by the desires of his financial backers. Hubbard, in particular, felt that the idea of a voice-transmittal device was ridiculous, a mere "scientific toy." He remained firmly convinced that a harmonic telegraph was of greater practical and commercial appeal. He felt that there was money to be made in the harmonic telegraph and that there were few prospects other than failure and ridicule likely for Bell's new idea.

Hubbard strongly urged—perhaps *bullied*—Bell to stay the course. Both Hubbard and Sanders threatened to suspend their funding if Bell persisted in work on the telephone rather than the harmonic telegraph. Hubbard vowed to obstruct Bell's engagement to his daughter Mabel and even demanded that Mabel tell Bell that she would not marry him unless he completed work on the harmonic telegraph first.

Rapidly developing events—including lucky accidents and conceptual breakthroughs—were soon to take over, however, and shove Hubbard's objections to the side. This cleared the way for Bell and Watson's creation of a working telephone.

A HAPPY ACCIDENT

Until this point, all telegraphs, including Gray's and Bell's competing prototypes for the harmonic telegraph, were based on the principle of intermittent electrical current. The controlled disruption of the electrical current running through the wires created the electrical pulses that conveyed the coded content of transmitted messages. Bell and Watson were about to stumble on a powerful and innovative alternative, one that would enable the human voice to be carried across wires and reproduced faithfully on the receiving end.

Only a month before the lucky mishap that occurred in Bell's workshop in July 1875—a mishap that would lead directly to the world's first working telephone—Bell had a vague notion of the principle behind his eventual breakthrough. In May, he wrote to Hubbard (according to Shulman): "I have read somewhere that the resistance offered by a wire . . . is affected by the tension of the wire. If this is so, a continuous current of electricity passed through a vibrating wire should meet with a varying resistance, and hence a pulsatory action should be induced in the current."

What this means is that a controlled interruption of the electrical current—the repeated full opening and absolute closing of an electrical circuit—is not necessary to transmit sound. Rather, a continuous electrical current passing through a vibrating wire will generate enough pulsation to convey not only a sound corresponding to whatever electromagnetic process had generated it but also the accurate timbre, tone, and texture of the transmitted sound itself. This is known as an undulating rather than intermittent current, and it more closely resembles the movement of the human voice, its fluid ups and downs of tone and volume. An undulating current echoes the patterns of actual sound waves. There would seem to be no obstacle to the electrical transmission of the human voice. Bell was close to the conceptual breakthrough he needed to be able to create a working telephone. The equally necessary practical, mechanical breakthrough would occur next, with the invaluable assistance of chance, and confirm Bell's still-emerging theory about the utility of continuous currents of electricity.

On the afternoon of June 2, 1875, Bell and Watson were dutifully working on the harmonic telegraph at Charles Williams's machine shop, trying to perfect the invention that was no longer their primary interest but that their investors still wanted to be completed, patented, and marketed. Watson manned the transmitting apparatus in one room while Bell sat at the receiver in an adjoining room. Watson was striking the keys that in turn struck the transmitting reeds, but one of the reeds was not responding to the keystroke and failed to emit any tone.

Out of sheer frustration with the device, which seemed to fail more often than not, Watson began to pluck the troublesome reed. Almost immediately, Bell burst into the room demanding to know what Watson had done and ordering him to do it again. Bell had distinctly heard the plucked reed through his receiver. He had not heard it just through the office walls, and he had not heard just the tone of the plucked reed. He had actually heard the plucking of the reed *and* the resulting tone. Both sounds had been transmitted along the wire that ran between the two rooms and reproduced audibly and accurately through his receiver.

The two men soon discovered that the stuck reed had been pinned in place by a contact screw that had been overtightened, which caused the electrical current to become continuous and steady rather than intermittent. The continuous electrical current could carry not only the tones generated by the electromagnetic forces unleashed by the magnetized reeds and electric wire but all of the sound waves generated by the plucking of the reed and the reed's corresponding tone. This one brief, accidental transmission proved that any sound, including that of the human voice, could be transmitted electrically and reproduced in all its richness of tones, timbres, and pitches. No longer would clicks or mere musical tones be the only sounds conveyed over wires.

A FRUSTRATING LULL

Bell and Watson spent the remainder of the day and much of the night trying without success to duplicate this accidental result. Bell also began to sketch a simple telephone transmitter and receiver system that he wanted Watson to begin building immediately and have ready for experimentation by the following night. The design would use aspects of the harmonic telegraph but would incorporate the discoveries Bell had made with his "autograph telegraph" experiments and the recent stuck-reed mishap.

Watson attached one of the magnetized harmonic reeds to a simple wood frame. One end of the reed was attached to a thin parchment that would work like the delicate membrane of the human eardrum. The parchment was attached to a wire that ran to the telephone receiver. At the receiving end, the wire made contact with a magnetized reed, which in turn was attached to another piece of parchment. A speaker at the transmitting end would speak with his lips very close to the parchment, which would vibrate as a result. These vibrations would cause the magnetized reed to vibrate accordingly. The reed's vibrations would generate an electrical current, which would then speed through the wire. When the current hit the reed on the receiving end, the reed would vibrate and cause the receiving-end parchment to vibrate as well. A listener whose ear was very close to the parchment would, in theory, hear the original sound clearly and accurately, in the very instant that it was spoken at the transmitter end.

Watson was as good as his word and had the prototype telephones ready by the following evening. They tested it repeatedly, with mostly unsuccessful results. Watson claimed to be able to hear the tone of Bell's voice through the receiver and the occasional isolated word. Bell ended up shouting through the transmitter so loudly and violently that he actually burst the parchment. The night's results were not clear enough for the two men to be able to declare success and announce their achievement.

Indeed, it would be another nine months before Bell and Watson would achieve a definitive breakthrough that allowed them to declare beyond a shadow of a doubt that they had created a working telephone. In the meantime, Bell slid into a depression as a result of his inability to discover which piece of the puzzle he was missing in order to perfect his invention. At this time, he wrote to his parents of his frustration and stubborn optimism (as quoted by Shulman), "I am like a man in a fog who is sure of his latitude and longitude. I know that I am close to the land for which I am bound and when the fog lifts I shall see it right before me."

Bell's despair and frustration were worsened by Hubbard's increasing impatience with his lack of progress on the harmonic telegraph and his growing distraction with the seemingly foolish telephone experiments. Hubbard again threatened Bell with denying him his daughter's hand in marriage if he did not return to the telegraph project. This was a grave miscalculation on Hubbard's part. Bell, who ordinarily was meek and agreeable with his financial backers, was at this point so inflamed by both his love for Mabel and his proximity to a telephone breakthrough that he abandoned his usual reserve and fought back with great force and passion. Against Hubbard's stated demands, Bell and Mabel announced their engagement, following a sharp and angry argument between the two men.

THE "EUREKA" MOMENT

Hubbard and Bell were able to patch things up well enough that Bell's future father-in-law helped him to file his patent application for the telephone in February 1876. Bell had worked throughout the fall and winter on his patent specifications even though he still had not created a true working model of the invention they described. Perhaps it was therapeutic to him to describe

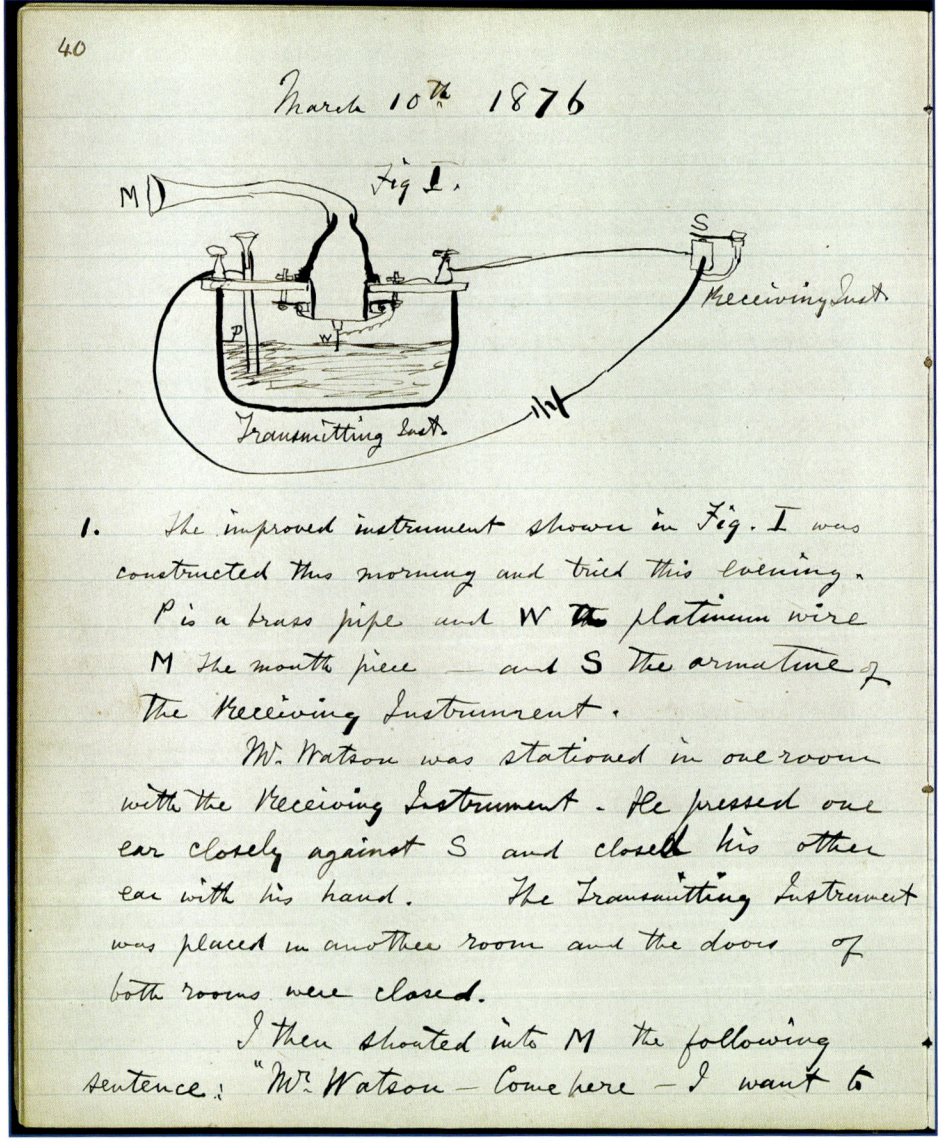

40

March 10th 1876

Fig I.

M

Receiving Inst.

P   W

Transmitting Inst.

1.   The improved instrument shown in Fig. I was constructed this morning and tried this evening. P is a brass pipe and W the platinum wire M the mouth piece — and S the armature of the Receiving Instrument.
       Mr. Watson was stationed in one room with the Receiving Instrument. He pressed one ear closely against S and closed his other ear with his hand.   The Transmitting Instrument was placed in another room and the doors of both rooms were closed.
       I then shouted into M the following sentence: "Mr. Watson — Come here — I want to

Alexander Graham Bell continued with his work despite lack of support from his patrons. It would take Bell and Watson several years of trial and error before they developed a successful prototype. In this page from Bell's lab notebook, a sketch of a device with a tub of acid is included with the day's observations. Dated March 10, 1876, this entry details the first successful conversation transmitted over the telephone.

in writing the principles and mechanics of an ideally working machine while he tried and failed to make the actual invention perform as designed. Perhaps, too, this provided the necessary moment of inspiration—the proverbial "aha" or "eureka" moment—that allowed Bell to finally figure out how to clearly transmit speech over wires and close the gap between vision and reality, design and utility, plan and execution.

This "eureka" moment apparently arrived on March 8, 1876. All of Bell's previous telephone experiments had used electromagnets and the vibrations generated by the human voice itself to generate the current through the telephone wire that would, theoretically, carry sound to the receiver on the other end. Quite abruptly, Bell abandoned electromagnets and began to focus instead on using shallow dishes containing diluted sulfuric acid to generate an electrical current. In doing so, Bell had apparently struck upon the principle of variable resistance.

In their earlier breakthrough in June 1875, Bell and Watson had discovered that a continuous electrical current conveyed sound waves in a way that an interrupted electrical current could not. This is why a conventional telegraph could transmit only electric pulse-induced clicks, not the human voice or any other true sound generated at a transmitter. Now, however, they had struck upon the realization that the current had to be steady but variable. It should not be absolutely opened and then closed, as with the telegraph, but neither should it remain entirely open at all times. If the current and its power could fluctuate in exact accordance with the sound waves being transmitted—gaining power with an increase in volume, reducing power with lower volume or silences—then sound could be transmitted with greater clarity over greater distances. The electrical current must pulse in exactly the same way as the sound wave to be transmitted. The problem with Bell and Watson's previous attempts at transmitting sound was that the electromagnets could generate only enough power to carry sound over short distances, and even these results were not as clear as they hoped. Unlike the magneto-induction caused by the proximity of a membrane to an electromagnet, voice-generated variable resistance resulted in long-range amplification of sound—not just the conveyance of sound, which became increasingly feeble with distance.

With Bell's direction, Watson again set to work making a prototype that used this new concept. The transmitter now featured a wire attached to a diaphragm, a tubelike mouthpiece. The wire's other end rested in a small basin of diluted sulfuric acid. When someone spoke into the diaphragm, the wire would vibrate and sink into the acid solution. The louder and more resonant the voice, the more the wire would sink. This fluctuation in the wire's movements would generate a similarly variable electrical current in a battery that was wired to the receiver. The receiver wire would then vibrate, which caused the receiver's membrane to vibrate as well. The membrane's vibrations would generate sound, faithfully and instantly reproducing the very same sounds produced at the transmitter's diaphragm. The deeper the wire dipped into the solution, the less resistance the liquid offered to the electrical current, which allowed for greater clarity and distance of transmission.

In theory, this sounded logical to Bell and Watson, but the proof would lie in the practical application of the theory. It had been a long time since the two men had tasted success in their experiments in the Boston workshop. Fateful chance and happy accidents had nudged their work forward several times in the past, but luck seemed to have abandoned them. It was about to return, just when they needed it most and were in danger of losing the race to invent the telephone.

THE FIRST TELEPHONE CALL

On March 10, 1876, only two days after Bell appeared to shift his attention to variable resistance experiments, he and Watson shared the world's very first telephone conversation. Appropriately, it was achieved partly by design and partly by accident. Watson, who had just set up the new transmitter and receiver in two rooms of their new Boston workrooms on Exeter Place, stood by the receiving telephone in one room and awaited Bell's hoped-for voice transmission from an adjoining room. In the other room, Bell spilled sulfuric acid on himself while settling before the transmitter and instinctively shouted into the diaphragm, "Mr. Watson! Come here. I want you."

Watson heard Bell's voice loud and clear and even picked up on the urgency of his tone—he correctly assumed that Bell was distressed. Immediately, he abandoned the receiver and rushed to Bell's workroom to see what was the matter. When Watson arrived, Bell realized that his assistant must have responded to his voice transmission because he would not have been heard any other way. Watson quickly affirmed that he had heard Bell's words clearly and distinctly. According to John Brooks, Watson reported that Bell "forgot the [acid] accident in his joy over the success of the new transmitter."

Bell had every reason to be overjoyed and indifferent to discomfort. He had become the first person to create a working telephone and could stake claim to being the inventor of this innovative new technology. This claim would not go undisputed, however, and would be bitterly contested for many years afterward. Indeed, the 1876 events surrounding Bell's patent applications and decisive "eureka" moment remain clouded by uncertainty and troubling accusations of deceit, intellectual thievery, and cheating.

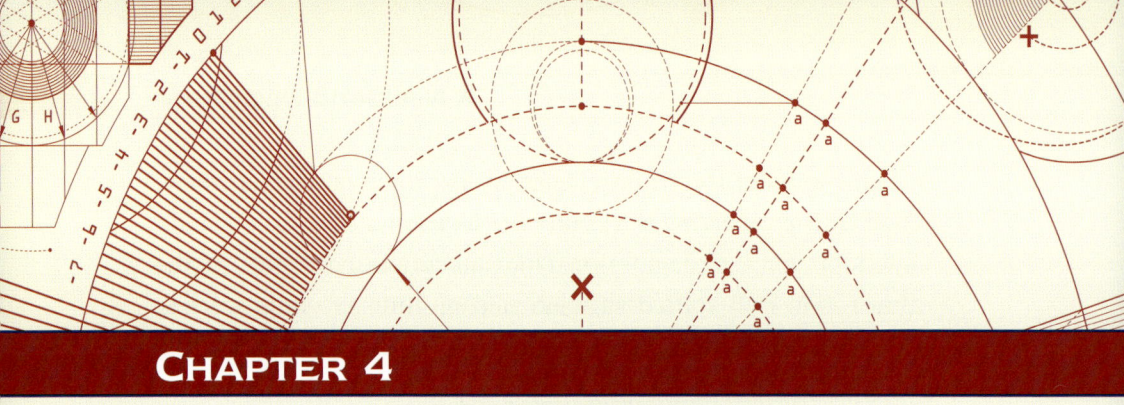

CHAPTER 4

A Cloud of Suspicion

Alexander Graham Bell had filed his patent application for the telephone on February 14, 1876, more than three weeks before his first successful telephone transmission. According to most accounts, it was his primary financial backer, Gardiner Greene Hubbard, who hand delivered the application to the Patent Office in Washington, D.C., while, presumably, Bell continued to labor in Boston toward the breakthrough that would finally allow him to perfect a working apparatus.

A PUZZLING PATENT APPLICATION

Yet, there are several curious facts associated with this filing. The first is that, in Bell's patent application (which he titled with a strange reserve and lack of fanfare "Improvements in Telegraphy"), he made little mention of the telephone's primary, and most revolutionary, function—the ability to transmit speech audibly and clearly. Instead, four of the five claims in his patent application were concerned more with the technology behind his harmonic telegraph. It described an apparatus that generates a "vibratory or undulatory current of electricity" in order to transmit and receive several telegraphic messages over the same wire simultaneously and without interfering with one another. The messages themselves were described more as signals, primarily musical notes of varying volume and pitch and other "noises or sounds of any kind."

Buried at the bottom of these claims for an undeniably innovative but hardly earth-shaking invention was a fifth and final claim that by all rights should have topped the list and dominated the patent application's content. Here, in bland and understated language—the written equivalent of a mumble—Bell introduced the utterly transformative concept of variable resistance and telephony, quietly announcing his invention of an apparatus that could transmit "vocal or other sounds telegraphically" with the help of electrical "undulations" that closely resemble the sound waves that generate the current in the first place. In effect, then, Bell described not one but two inventions in this patent application—the harmonic telegraph and the telephone (though the word *telephone* was never mentioned in the document). For whatever reason, the less important invention was given the most attention.

The only problem—and the second curious fact about Bell's patent application—is that Bell had not yet invented the telephone he described in the document. Not only had he not successfully tested a properly working model of the described apparatus, but—by the date of the application's filing—he had not even begun his experiments on variable resistance (as established by the timeline laid out in his and Watson's journal entries). How could Bell have solved the central problem that prevented him from creating a working telephone but not yet have conducted the necessary experiments that would have provided the solution? If he had solved the problem in theory, why had he not yet put the theory to a real-world test? Why would he confidently include the theory in his application, presented as established fact, before being certain that he was right and could definitely produce the working apparatus he described?

The mystery deepens when the actual handwritten draft of Bell's application is examined. The bulk of it, including the four claims based on the harmonic telegraph technology, appears to have been carefully considered, composed, and written out. Yet toward the end of the document, the fifth claim—which concerns the theory and practice of variable resistance and telephone technology—is closely scrawled perpendicular to the main text and in the paper's margin. A thin, wavering line connects it to the place in the text where it should be situated. According to Bell, he wrote this complete draft—including the marginal insertion—in early January 1876 after three months of revising the application material. He completed the document on January 12 and sent it to a Washington, D.C., law office the next day to be copied and prepared for his signature and submission to the patent office. Despite the fact that the content contained within the variable resistance paragraph is of central importance to Bell's telephone apparatus, he claimed that he almost sent off the application without including it. Realizing at the last minute his absent-minded omission, he quickly scribbled the crucial paragraph in the margin, indicated where it should be placed, and sent the draft.

All of this confusion could be chalked up to the typically chaotic but brilliant minds of ingenious inventors who master complicated principles but are easily overwhelmed by the ordinary activities of everyday life. After all, Bell's first telephone call relayed a message of distress about spilled sulfuric acid, and his first inkling of the possibility of transmitted speech was inspired by a mistranslation of German. Bell's story could be accepted at face value and could simply enhance the sweet and charming image of a somewhat bumbling inventor who started out by trying to improve the lives of his deaf pupils, was as devoted to his beloved

Alexander Graham Bell's patent application for the telephone was filed just hours before Elisha Gray, Bell's rival, filed his own caveat for a similar machine. Despite the patent controversy, it was Bell's device that became popular and successful. *Above*, a twentieth-century couple demonstrates Bell's patent model of the telephone.

Mabel as he was to his groundbreaking work, and broke into victory dances with his assistant, Watson, to celebrate successful experiments. Yet, there exists a significant obstacle to accepting Bell's story without reservation—his old rival Elisha Gray.

Bell once described their race to invent the harmonic telegraph as "neck-and-neck." It turned out that they were actually in a far more important race to invent the telephone. It was not just neck and neck, it was a virtual dead heat—a photo finish.

A HURRIED FILING

Bell's application would not be filed with the patent office until February 14 because a copy was first entrusted to a Canadian family friend, George Brown, to submit to English patent officials. England only accepted patent applications for inventions that had not yet received patents in any other countries. Bell, Hubbard, and Sanders had decided to hold off on submitting Bell's application to the U.S. Patent Office until they heard back from English officials. At some point, and for unknown reasons, they abandoned this strategy and abruptly filed the application in Washington.

In what may have been one of history's curious, one-in-a-million coincidences, Bell filed his patent application on the very same day as Gray submitted a caveat for a machine that could transmit and receive vocal sounds and entire conversations telegraphically and long-distance via an electric circuit. A caveat was distinct from a patent. Receiving a caveat allowed an inventor to explain an idea for an invention for which he or she had not yet been able to produce a working model. The idea was protected for a year, at the end of which period the inventor had to produce a working model or lose the protection. At that point, any competitor could come along and poach the idea, use it to create an invention, patent it, and receive all of the profits derived from that patent.

Bell's patent application and Gray's caveat were filed within hours of each other. Later, each inventor would claim that his application had arrived at the patent office first and should therefore be considered the first official declaration of the invention of telephone technology. Bell's representatives claimed that his application was brought to the patent office in the morning. They added that its deliverer insisted that the application be officially noted and hand delivered to the patent examiner's office. These actions would seem to be unnecessarily urgent unless they knew that a competitor was also on the brink of filing a similar patent application or caveat. Bell's camp also claimed that Gray's caveat arrived at the patent office two hours later or sometime in the afternoon and may not have reached the examiner until the next day. Gray countered that he remembered his caveat being filed in the morning. Historians have presented various accounts and evidence to suggest the truth or falsity of one or another version of this "he said/he said" story.

Patent officials did not stamp incoming applications with the time of their arrival, so the hour of the submission did not really matter. Shulman notes that patent examiners were concerned with who could prove that they had invented a given device first or who had conceived of it first, not who could be the first to file an application claiming invention. In this unusual case of a near-simultaneous filing of a patent application and a caveat for, in effect, the very same invention, patent officials should have suspended Bell's patent application for three months and offered Gray the chance to use this time to file his own patent application instead of the caveat. Once both applications were in hand, the merits of each would be considered, as would whatever working devices either man could produce to prove his theory.

This standard operating procedure was not followed, however. Instead, a number of irregularities and lapses occurred, resulting in the rapid and unequivocal awarding to Bell of an exclusive patent on the telephone. To what degree these lapses and deviations from official procedure were honest and accidental, and to what degree they were influenced by Bell's powerful friends and financial backers, is the troubling question that has hung over the invention of the telephone from day one and continues to cloud Bell's legacy.

AN UNDENIABLE ACHIEVEMENT

The tale of what may or may not have occurred within the patent offices during the last weeks of February 1876 is too complex to tell here. The strange and troubling saga is dramatically and thoroughly recounted by Seth Shulman in his book *The Telephone Gambit: Chasing Alexander Graham Bell's Secret.*

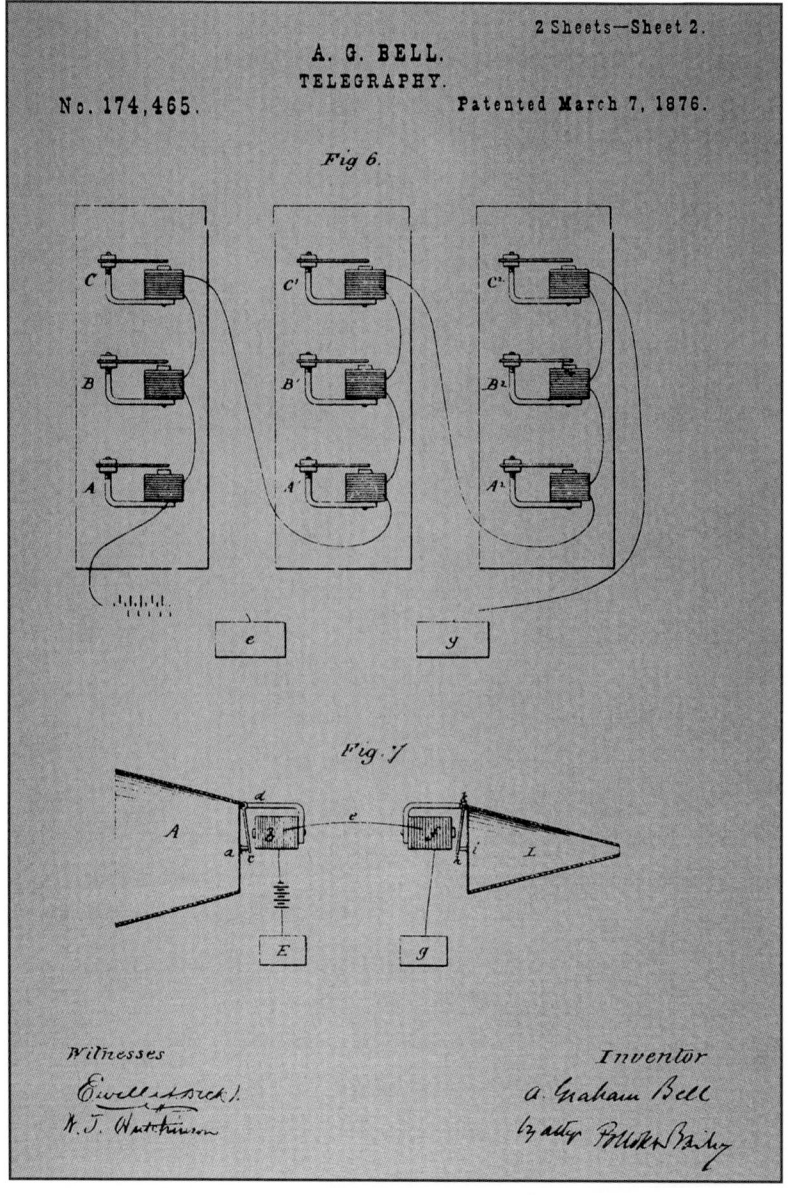

Claims that Alexander Graham Bell had altered the designs in his patent application *(above)* to include information from Elisha Gray's caveat proposal cast a shadow over Bell's success. Critics later agreed that Bell was the rightful inventor of the telephone because he had a working model of his machine, while it took Gray some time to develop speech transmissions.

Shulman argues that Bell stole Gray's ideas on variable resistance and the liquid transmitter while perusing his caveat in the patent office, which was a serious violation of the rules. He then doctored his own patent application by inserting Gray's ideas and rushed back to Boston to test them.

Yet the fact remains that, although Elisha Gray may have been the first to employ variable resistance as the crucial impetus to transmitted speech, he had been unable to build a working apparatus, even in the months after the patent controversy erupted. In fact, in a letter to Bell written on March 5, 1877, requesting permission to publicly demonstrate Bell's telephone apparatus, Gray stated (according to Brooks), "I do not . . . claim even the credit for inventing [the telephone], as I do not believe a mere description of an idea that has never been *reduced to practice* . . . should be dignified with the name invention." Though Gray would later deny Bell's claim to being first, the truth inherent in this unguarded statement is difficult to argue with—Bell may not have had the idea for the telephone or its engineering requirements first, but he put these ideas into practical, working action before anyone else did. He filed a full patent application, whereas Gray merely submitted a caveat. Bell, apparently, was ready to build his telephone apparatus; Gray, in filing a caveat, declared that he was close but not there yet. Therein lies the crucial, defining difference between their two fates and legacies.

It is this undeniable achievement, this historical first, that helps cut through the dense fog of mystery surrounding Alexander Graham Bell's patent application and final days of experimentation before his first successful telephone transmission on March 10, 1876. Too many unanswered questions remain about Bell's supposed intellectual thievery to convict him in the court of general opinion.

In the end, all that can be said is that Bell had assimilated the efforts and innovations of prior inventors (as all inventors do, whether they acknowledge it or not); had introduced his own innovations, which resulted in the new harmonic telegraph; and had been devoted to his work on the telephone for several years, despite the discouragement of his backers. He had stubbornly soldiered on and, somehow, arrived at the crucial operating principle of variable resistance and constructed a working apparatus before Elisha Gray, whose work closely paralleled his. (This is often the case just before a scientific breakthrough; such breakthroughs are often the result of a very close race between a group of fiercely competing researchers, scientists, and inventors.) Even Shulman feels obliged to note that, "Bell may well have stolen Gray's design for the breakthrough variable resistance transmitter, yet there is little doubt that Gray, locked as he was into his shorter-term interest in the telegraph industry's sought-after multiple telegraph, would probably have been slow to commercialize the telephone even if he did pioneer it . . . Bell's vision and his energy still stand out as a remarkable model."

As John Brooks concludes, "But this should be clear: underneath the legal intricacies and the suspicions of foul play, it appears that by any rational standard Bell actually invented the telephone; Gray had not transmitted speech at the time of the [patent and caveat] filings and would not do so for many months afterward . . . No matter, the telephone was born at last." What we are left with, quite simply, is Bell and his working apparatus and the first clearly understood telephone call ever to be placed and received in human history. Alexander Graham Bell had gotten there first, as was established and reestablished in the almost decade's worth of court proceedings and more than 600 lawsuits that would follow the awarding of his patent. His claim, though fiercely questioned in his own day and still strongly challenged at the dawn of the twenty-first century, remains strong. In the eyes of history, Alexander Graham Bell—with the help of his assistant, Thomas A. Watson—was, is, and will remain the inventor of the telephone.

CHAPTER 5

Selling the Telephone

Alexander Graham Bell was awarded his patent for the telephone on March 7, 1876. Though it (and more of his telephone patents) would be contested in court until 1888, Bell, Hubbard, and Sanders could now begin to publicize the invention, market it, and try to build telephone networks throughout the country. The first crucial step in this long process was to introduce Americans to the telephone and begin building excitement and demand for it.

THE INTERNATIONAL CENTENNIAL EXPOSITION

Bell began immediately by offering a couple of lectures and demonstrations in Boston to scientific clubs and organizations in May 1876. An opportunity to greatly expand the reach of his demonstrations presented itself only a month later. In June, in honor of the nation's one hundredth birthday, Philadelphia hosted an international fair called the International Centennial Exposition. Like most World's Fairs, the popular focus of the

Centennial Exposition would be on those exhibits that pointed to the technology of the future, including the first electric light, Western Union's new printing telegraphs, and Elisha Gray's multiple telegraph. Even the arm and torch of the future Statue of Liberty debuted at the Centennial Exposition.

The assembled judges and onlookers included some of the day's greatest scientific luminaries, such as Lord Kelvin and Joseph Henry, whom Bell had consulted about both the harmonic telegraph and the telephone in 1875. After a brief explanation by Bell of the basic mechanics and principles, Dom Pedro II, emperor of Brazil, walked over to the receiver, put his ear to it, and listened as Bell delivered Hamlet's "To be, or not to be" soliloquy into the transmitter's diaphragm. By some accounts, Dom Pedro exclaimed in Portuguese, "My God! It talks!" According to still other versions of the story, he declared, "I hear! I hear!"

Joseph Henry and Lord Kelvin both took turns at the receiver next. Though they were extremely accomplished, experienced, and knowledgeable men of science, seemingly beyond mystification and wonder, both were astonished not by what they heard but by the fact that they heard anything at all. Both men had an intimate and thorough understanding of electromagnetism, telegraphy, and sound transmission, but both were awestruck by Bell's invention. At the time, Lord Kelvin seconded Dom Pedro's shocked reaction, affirming, "It DOES speak! It is the most wonderful thing I have seen in America." In his judge's report on offering Bell a Centennial Certificate of Award, Lord Kelvin further stated (according to Casson), "Mr. Bell has achieved a result of transcendent scientific interest. I heard it speak distinctly several sentences . . . I was astonished and delighted . . . It is the greatest marvel hitherto achieved by the electric telegraph."

These stunning results were achieved by Bell's magneto transmitter. In the end, he did not use and had no apparent need for the liquid transmitter that caused so much controversy. Bell achieved his enormous success at the Centennial Exposition without it.

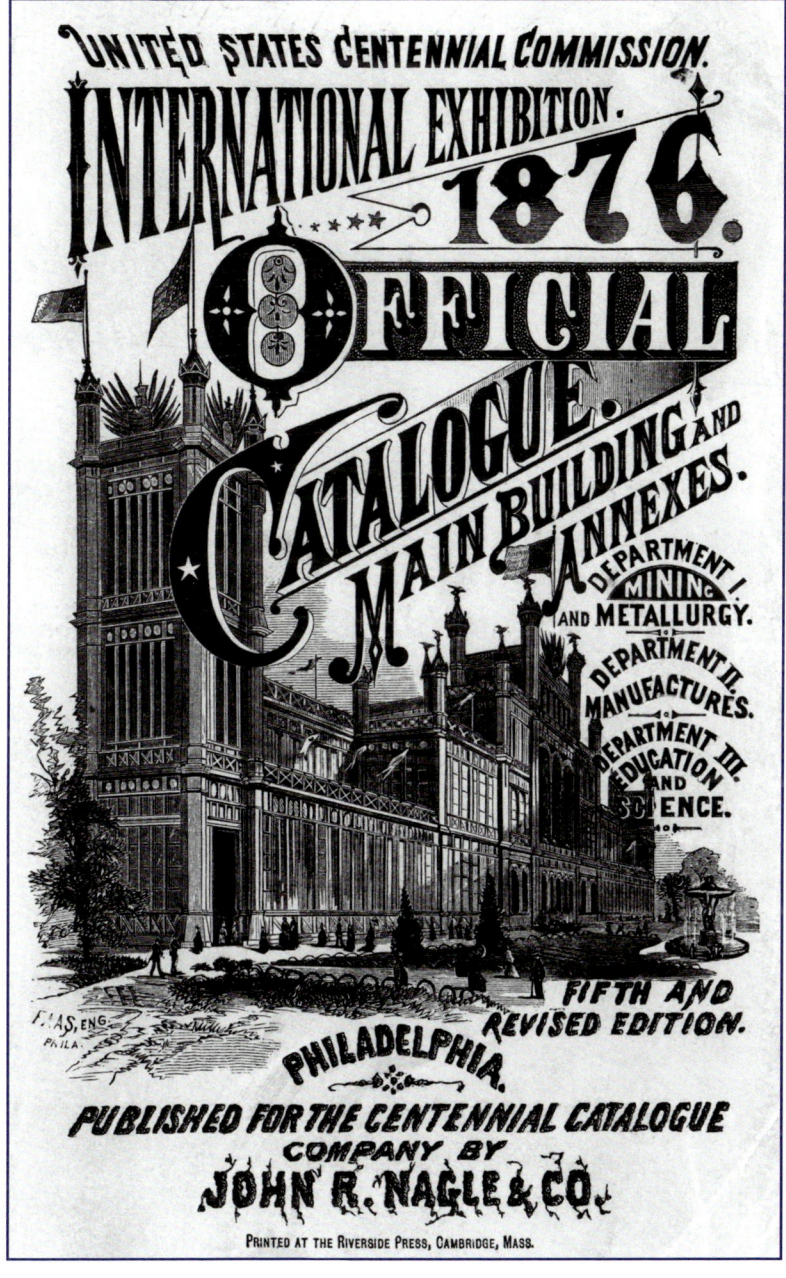

The telephone was first unveiled at the 1876 Centennial Exposition in Philadelphia. The successful transmission of sound shocked the audience, which included important scientists and businessmen. *Above*, the cover of the 1876 Centennial Exposition catalogue.

Although the scientific experts and dignitaries who witnessed Bell's demonstration of his telephone that day were overwhelmed with wonder for this "talking telegraph," many ordinary people who tried it out throughout the Centennial's summer-long exhibit were less impressed. Many felt that it was a novelty at best, with no conceivable practical value. Some felt awkward speaking into the diaphragm, as if they were performing rather than holding a natural, easy conversation. Some people disliked having to shout into the diaphragm to be heard clearly at the receiver end. Others were convinced that it was all mere trickery, a hoax, and that someone was hiding behind a curtain or door, throwing his voice. There were even some who suspected witchcraft or supernatural manipulation. Some people mocked it as a parlor trick, and others simply dismissed it as a novelty toy.

These reactions began to be mirrored by the press, business interests, and governments. As Bell and his backers discovered—after they formed a company to market the device and show it off in carnival attraction–like demonstrations—the vast majority of Americans could not wrap their minds around the paradigm shift that a telephone conversation represented. To them, as to Elisha Gray and Gardiner Greene Hubbard and many other scientifically knowledgeable and influential decision-makers, the multiple telegraph was a far more immediately useful and exciting technological advance. Only those few individuals blessed with the gift of foresight—such as Lord Kelvin, Joseph Henry, and Bell himself—saw how the telephone would revolutionize communications and soon render the telegraph an antique.

Following the uproar—both enthusiastic and mocking—generated at the International Centennial Exposition, Bell, Watson, Hubbard, and Sanders sought to improve the telephone technology and its smooth, clear functioning and sell it to a largely skeptical and dismissive nation. Between August 1876 and April 1877, Bell and Watson conducted several important and mostly successful tests of their new magneto transmitter and receiver over increasingly long distances. Borrowing existing telegraph wires, they managed to hold intelligible phone conversations between Brantford and Paris, Ontario, in August. This was the very first telephone call in history that was conveyed over outdoor wires. This encouraging success was followed in the months ahead by similar calls placed between Boston and Cambridge, Massachusetts, and between Boston and New York City.

GROWING INTEREST AND ENTHUSIASM

In the meantime, in February 1877, Bell's invention became even more secure from the attacks of his rivals when the patent office issued him a second patent for the actual transmitter, receiver, and mechanical operations of the telephone (rather than its theoretical scientific principles protected in the first patent). A series of important "firsts" quickly followed. A telephone was first used to report news in February 1877, when an account of one of Bell's lecture-demonstrations in Salem, Massachusetts, was telephoned in to a reporter from the *Boston Globe*. The world's first permanent outdoor phone line was strung between Charles Williams's electrical workshop in Boston (where Watson and Bell had labored so hard on the harmonic telegraph and telephone in the early days) and his home in Somerville, Massachusetts. Another Boston-to-Somerville phone line was created to run from a banker's office to his home, marking the first time that telephone equipment and service were rented for business use.

The growing technical successes and early indications of interest among businesspeople for telephone service probably helped to reassure a dubious Hubbard about the wisdom of throwing his money and full energies behind an instrument whose usefulness he still did not entirely understand. In fact, in late 1876 or early 1877, Hubbard actually tried to cut his investment losses and recoup some part of them by offering to sell Bell's telephone patent to his arch-nemesis Western Union for $100,000. Either the powers-that-be at Western Union also failed

In an effort to raise awareness about the significance of the invention and also to recoup some investment losses, Gardiner Greene Hubbard sent Thomas Watson and Alexander Graham Bell on a speaking tour to demonstrate the new machine *(above)*.

to recognize the telephone's enormous potential (the company's president had dismissed it as a useless "electrical toy") or they disliked and distrusted Hubbard too much, given his constant attempts to break up their telegraph monopoly. Whatever the case, Hubbard's offer was rejected.

With increasing technical success and the first spark of commercial interest, Hubbard sent Bell and Watson out on the road to demonstrate, publicize, and market their invention. They traveled throughout the Northeast, renting out lecture halls and telegraph wires, and charging admission. Bell would man the telephone receiver in the lecture hall, while Watson and some musicians would be stationed at least several miles away with the transmitter. Bell and Watson would converse, the band would play music (including "The Star-Spangled Banner"), and Watson would recite Shakespeare and even sing popular songs, such as "Yankee Doodle Dandy" and "Auld Lang Syne."

Most of these demonstrations went well, and the audiences often included important businesspeople, politicians, scientists, and other influential men of society. Bell even traveled to England to demonstrate the telephone to Queen Victoria. At her request, he placed a permanent phone in Buckingham Palace and wired it to another phone at her royal home on the Isle of Wight. Excitement was starting to build, and Hubbard worked hard to stoke this excitement into actual demand for phone service. People had to be persuaded that they needed the telephone and not simply view it as a thrilling curiosity or penny arcade entertainment.

During his own business travels, Hubbard always carried a transmitter and receiver to demonstrate them to anyone who showed interest. He also began to advertise phone service and equipment rental, offering to create a private telephone line between two locations no more than 20 miles from each other. Usually one phone was placed in a businessman's office and wired to a second phone in his house. The service and rental fees cost $20 a year for private or "social" use and $40 a year for business use.

In addition, Hubbard was instrumental in creating the world's first telephone exchange—a network of multiple telephones connected together rather than a simple phone line between two phones. To get the word out about phones and build demand among businesspeople, he lent 12 phones free of charge to E.T. Holmes, a Boston man who owned a burglar alarm company. Holmes installed five of them in five local banks and wired them to a switch in his own office. This relatively simple exchange allowed someone in Holmes's office to connect the phone lines of the bankers. During business hours, the bankers could use these phones to communicate with each other.

BELL TELEPHONE IS BORN

Bell's entertaining demonstrations and Hubbard's entrepreneurial flair and doggedness began to pay off. By mid-1877, there were between 600 and 800 telephone customers with private phone-to-phone lines (known as "point-to-point" or "two-point" communications). Interest in the telephone was beginning to pick up as public skepticism, especially among businesspeople, gave way to an increasing awareness of the practical value of instantaneous, person-to-person spoken communication. Hubbard and Sanders, who had spent hundreds of thousands of dollars on Bell's experimentation, were now eager to recoup their investment and, hopefully, begin turning a profit.

On July 9, 1877, together with Bell and Watson, they formed the Bell Telephone Company. Sanders, Hubbard, and his daughter Mabel owned roughly equal shares; each held just under one-third of the private company's 5,000 shares. Watson was granted 499 shares. Bell, the telephone's inventor, received the least shares. Indeed, he received only 10 shares, the same number that Hubbard's brother was offered and fewer than Hubbard's wife. It would appear that this was how Bell wanted it. Initially, he was offered slightly more shares than Hubbard and Sanders but promptly turned most of them over to Mabel. Committed to experimentation and invention rather than finances, commerce, and profiteering, Bell was probably more than content with the achievement of the telephone itself. He did not need to become rich by it. He may have also felt that the men (and their families) who had supported him financially and emotionally during those long, frustrating years of invention should receive the lion's share of whatever profit could be made from his work.

For Bell, the work itself and its successful completion was enough. The real prize for him was Mabel Hubbard, whom he married two days after the formation of Bell Telephone. At this point, Alexander Graham Bell more or less disappears from the ongoing history of the telephone's development and global spread. The company that would wire America bore his name, but his crucial contribution to the effort had come to an end.

CHASING CAPITAL

Alexander Graham Bell had initially conceived of an ambitious plan to wire all of America with telephone lines, connecting every citizen in a web of instantaneous, electric, spoken communication. He envisioned central switching offices that would allow "direct communication between any two places in [a] city" and intercity telephone lines running through rural areas so that "a man in one part of the country [could] communicate by word of mouth with another in a different place" (as quoted by George David Smith in his book *The Anatomy of a Business Strategy: Bell, Western Electric, and the Origins of the American Telephone Industry*). It was also hoped that Bell Telephone could manufacture and sell its own telephone equipment, thus making money on both telephone service and its necessary hardware.

The problem, however, was that Bell's investors were almost completely tapped out and could not embark on such an ambitious venture. Sanders, in particular, had no more to invest. Quietly and from behind the scenes, he had provided 90 percent of the funds that went into Bell's invention, including the paying of

Bell's rent and Watson's salary. Both Hubbard and Sanders were hoping to sell the patent rights to the telephone to a larger company that could found a national telephone network.

Despite the increased interest in Bell's invention, however, no company came forward to buy the rights to the telephone and its technology. Sanders and Hubbard were forced to continue to recoup some of their money by establishing widely scattered two-point interoffice telephone lines and home-to-office lines.

BUILDING AMERICA NOW

TELEPHONE RATES AND CAPACITY

When the New York-to-Chicago long-distance telephone line opened in 1892, it could handle only one call at a time, at a cost of $9 for the first five minutes (which would equal almost $200 in 2008). The New York-to-San Francisco line completed in 1915 cost $20.70 ($441.87 in 2008) for the first three minutes. The transatlantic radio-wave telephone service that linked the United States and England in 1927 could only handle one call at a time, at a cost of $75 ($929.31 in 2008) for the first three minutes, and the transpacific service introduced seven years later cost $39 ($627.49 in 2008). The transatlantic cable that was laid on the ocean floor and connected North America to Europe in 1956 could handle 36 calls at a time and cost $12 ($95.12 in 2008) for the first three minutes. AT&T was essentially the sole long-distance telephone provider, and its Bell subsidiaries controlled most of the local telephone business, for which subscribers in the early years of the twentieth century paid anywhere from $1 to $3.50 ($21.78 to $76.22 in 2008) a month for basic service. Eventually, long-distance rates fluctuated according to the time of day and volume of traffic. Daytime weekday calls were far more expensive than nighttime and weekend calls.

Severely strapped for cash, they were forced to rely on outside manufacturers to make the telephone equipment. Initially, Charles Williams's workshop was the sole manufacturer, but other small companies were contracted as demand outstripped supply. Sanders and Hubbard also did not have the financial resources to establish and maintain phone services for their subscribers; that, too, had to be contracted out to service providers who licensed the right to operate telephone businesses.

Today, hundreds of service providers offer long-distance and local dialing plans for either landlines or cell phones, or both. Long-distance calling can cost as little as two to five cents per minute, any time of day, and many providers charge no monthly fees for basic service. For example, AT&T offers a fairly typical home phone plan in which, for $40 a month, the subscriber gets unlimited domestic calling—which means the subscriber can call anywhere within the United States, any time of the day, and use as many minutes as he or she wishes. It also offers a variety of wireless (cell phone) plans, charging between $40 per month for 450 minutes of calling and $80 per month for 1,350 minutes.

Fiber-optic cables allow for the transmission of not just the human voice but data and video as well. Ten billion digital bits per second can be transmitted via fiber-optic cables—the equivalent of 120,000 phone calls at a time. New standards are set to quadruple fiber-optic capacity to 40 gigabytes/second, or 40 billion digital bits/second. Capacity for cell phone calls depends on the particular cell in which a caller is located. Each cell, or region within a cellular network, provides for a maximum capacity of callers based on average demand. In general, each cell should be able to handle thousands of calls simultaneously. AT&T's first mobile phone service, begun in 1946, could handle only between one and two dozen calls at any given time within a single antenna region.

These telephone franchises strung the telephone lines, created point-to-point networks and exchanges, and established larger, multi-subscriber exchanges. In short, they did all the hard work of creating phone networks and infrastructure. Bell Telephone simply provided the telephones and the technical support and know-how. Traveling Bell agents would visit local entrepreneurs and give them technical manuals that described how to set up, maintain, and repair telephone equipment and exchanges. Bell Telephone made little or no money on this outsourced equipment or service (though eventually, after Hubbard and Sanders's time, the company would profit from its relationships with equipment manufacturers and local service providers).

The way Hubbard and Sanders did make a small amount of money was born of necessity but proved to be a stroke of genius that would one day enrich Bell Telephone (and, later, AT&T) beyond these two men's wildest dreams. Hubbard decided that, rather than sell phone equipment to subscribers, Bell Telephone would lease it. Customers would make payments for equipment during the entire period of their phone service subscription. This allowed for a steady and near-perpetual stream of income rather than a one-time sale and fluctuating, unpredictable periods of demand. As Claude S. Fischer points out in *America Calling*, this savvy arrangement would be akin to gas companies leasing stoves and furnaces to its customers or the electric company being the sole provider of lamps. If a customer wanted phone service, he or she would be required to pay continuously for both the service itself and its apparatus. Without the telephone apparatus, all customers would be left with was an unsightly wire strung on their house, extending to the world beyond but conveying no messages whatsoever.

A SLEEPING GIANT WAKES

This first tentative movement toward a telephone monopoly began to bear some fruit. In 1878, about 10,000 Bell telephones were being leased. Whatever breathing room this afforded Hubbard and Sanders, however, was increasingly stifled by a developing economic depression that dampened consumer and business spending, and also by Western Union's brazen and aggressive entrance into the telephone business.

As Bell Telephone began to enjoy some modest success in luring business customers away from the telegraph and toward its phone service, Western Union finally recognized the enormous value and potential of the so-called "electric toy." The massive telegraph monopoly quickly assembled a dream team of electrical engineers, including Elisha Gray, Amos E. Dolbear (who, in 1876, had sketched out an alleged improvement on Bell's telephone, one that he felt entitled him to a share of any profits accrued from Bell's patents), and a young Thomas Alva Edison, who by this point in his career had already invented an electric pen and a telegraph that could be used to count cast votes. Arguing that Gray was the true inventor of the telephone, Western Union simply ignored Bell's patent protections and began to manufacture its own telephones. It established service along its existing and extensive network of telegraph wires and infrastructure, reaching customers in virtually every U.S. city, town, and small village.

Unfortunately for Hubbard and Sanders, Western Union—in addition to developing entirely new telephone subscribers—also found plenty of customers who were willing to switch their service from Bell Telephone. Charles Williams's workshop was overwhelmed by orders for the telephone apparatus. The phones took a long time to build and ship, and Williams could not keep up with the demand. The phones themselves were difficult to use. The apparatus that Watson had designed for commercial use featured a one-piece transmitter-receiver. The speaker had to shout into the device's diaphragm and then quickly place his or her ear next to the opening to catch the response of the person on the other end. Because Bell had switched to the less powerful magneto transmitter rather than a liquid transmitter (perhaps to avoid further charges of having stolen Gray's idea),

In order to distance his invention from Elisha Gray's telephone, Alexander Graham Bell created a new magneto phone *(above)* for commercial use. Using this new telephone, which was designed by Thomas Watson, was a nuisance, as users had to repeatedly shout into the machine's diaphragm. Though rudimentary, the demand for telephones rose.

the signal was comparatively weak and grew even weaker across longer distances. Callers often had to raise their voices and repeat themselves frequently to be heard and understood. Loud and distracting static along the wire was also a major problem. When the two speakers talked simultaneously, a muddle of sound would result. As Bell advertisements urged, in one of the first instances of an emerging telephone etiquette, customers had to listen when they were not speaking. Even if speaking over each other was a typical feature of "real" face-to-face conversation, it was inadvisable on a Bell telephone.

The time was ripe for a bigger, wealthier, and more sophisticated company to come along and usher telephone technology to the next developmental level—and make a fortune in the process. This is exactly what Western Union planned to do when, in late 1877, it formed a subsidiary called the American Speaking Telephone Company.

Western Union and its impressive team of engineers entered the telephone business with vigor and skill. They claimed to be violating no patent protection laws because the company believed it possessed the "only original telephone" invented by Gray, who, along with Edison and Dolbear—the so-called trio of "original inventors" of the device—was now designing and supplying new and improved telephones that were far superior to the crude, unreliable, and unloved Bell model.

Soon, Bell's point-to-point telephone connections were replaced by larger exchanges first created in northeastern cities like New Haven (where the first telephone exchange was established in January 1878), Hartford (linking a drugstore with several doctors), Boston, Philadelphia, and New York. Exchanges allowed telephone subscribers to call any other customers who subscribed to that exchange. These exchanges were mostly created by Western Union, which took advantage of its existing customer base and telegraph infrastructure. Western Union's telephone exchanges began to spread west to, among other places, upstate New York, Pennsylvania, Missouri, Illinois, Wisconsin, Iowa, Oregon, and California.

Bell Telephone tried to move into whatever region Western Union colonized and set up competing exchanges (which did not allow customers of one service to telephone customers of a competing service), but it was being outspent and outmaneuvered. If

(continues on page 64)

BUILDING AMERICA NOW

THE EVOLVING TELEPHONE

The first commercial telephone was produced in 1877. It was a humble and unassuming rectangular device that looked like a small wooden box or an old-fashioned pencil sharpener. A small opening at the front served as both the receiver and the transmitter. The electricity required to make it operate was generated by the vibrations of the human voice itself. It required the user to speak into the opening and then quickly place his or her ear there to listen for a response.

Later models were wall mounted and consisted of three separate compartments within a vertical wooden frame. The top compartment contained bells to alert people to incoming calls and a handle that, when cranked, transmitted an electrical signal to operators, letting them know that the user wished to place a call. Some later models used a push button instead of a crank (before the cranks and push buttons, callers had to pound on the transmitter or whistle sharply into it to get an operator's attention). The middle compartment contained a mounted transmitter and a handheld receiver strung to the side of the phone's wooden frame. A lower compartment of the frame held a battery. Wall-mounted phones became far more compact once "common" batteries came into use and telephones received their power from batteries placed in central exchanges.

In the 1890s, desk set phones were created. These were much smaller apparatuses that could rest on a desk. They were simply a transmitter mounted on a short, candlestick-like shaft and pedestal, with a separate handheld receiver that hung in a hooked cradle on the side of the phone column. The receiver was held to the ear in one hand

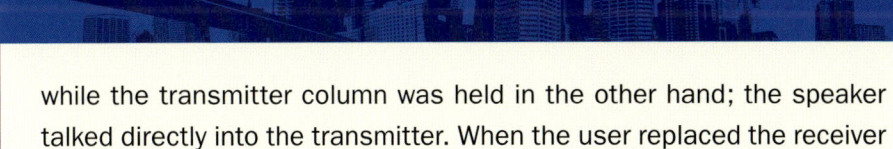

while the transmitter column was held in the other hand; the speaker talked directly into the transmitter. When the user replaced the receiver in its side-mounted cradle, the line was disconnected, signaling to the operator the end of the call.

Rotary-dial phones were first introduced in 1919, coinciding with the creation of the first dial telephone exchanges (rather than switches manned by operators). In the late 1920s, a new desk set was introduced that featured a handheld, one-piece receiver and transmitter, known as a handset. This handset was lifted off a telephone base and included a receiver end (placed at the ear) and a transmitter end (situated near the mouth). Based on designs already popular in Europe, this became known as the French phone.

In the late 1950s, AT&T began to offer its rotary-dial wall-mounted and desk phones in a variety of colors. In 1959, the popular Princess phone was first unveiled; it featured splashy colors, a slim base, and an illuminated rotary dial. Touch-tone push-button phones were first made available in 1964, and four years later the sleek and modular Trimline phone appeared, with its push buttons located on the handset rather than on the phone base.

The next great advance in phone design came with the introduction of cellular phone technology (first conceptualized by Bell Labs in the 1940s). Cell phones, which get smaller and more powerful every year, allow users to place calls to virtually anywhere in the world, from anywhere (in a car, on a train, or from a mountaintop). They can also act as digital cameras, MP3 players, day planners, address books, photo albums, camcorders, map libraries, mini-computers, and Internet portals. Telephones have come a long way from Bell and Watson's vibrating membranes, magnets, and metal wires.

(continued from page 61)

Bell Telephone was to survive and compete, it had to sue Western Union for patent infringement, which it did in September 1878.

REORGANIZATION AND NEW LEADERSHIP

Bell Telephone desperately needed an infusion of cash. Teetering on the edge of bankruptcy, its resources stretched thin, the company was stuck in a lengthy and costly series of lawsuits. At the same time, Bell continued to compete with Western Union for the wiring of America and the creation of nationwide telephone exchanges.

Thomas Sanders, stretched beyond his financial limit, called on various well-connected and wealthy relatives and their friends for investment dollars. These men responded favorably and supplied the cash by buying Bell stocks. In exchange, however, they demanded and received control over Bell Telephone. Their first decision was to reorganize the company to capitalize on its current strengths and expand its influence beyond this solid core. Because Bell's telephone had been born and developed in Boston, the company had gained a secure foothold in New England. To maintain this local dominance while making inroads elsewhere throughout the nation, Bell Telephone was split into two companies—the New England Telephone Company and a new Bell Telephone Company. The former was responsible for granting licenses to New England entrepreneurs who would establish local phone service using Bell equipment. The latter, based in New York City, would focus on issuing licenses to local phone companies throughout the rest of the country.

Bell Telephone's new investors received two votes per share of Bell stock owned, as opposed to the one vote per share enjoyed by Hubbard, Sanders, Bell, and Watson. Hubbard and Sanders had secured the present financial health and future prosperity of their small and still fragile company, but now they were removed from any real control over it. Hubbard's final action as trustee of Bell Telephone was to hire Theodore N. Vail as general manager of the company. As a member of the Congressional Postal Committee, Hubbard had become familiar with Vail, who was superintendent of the Railway Mail Service, a division of the U.S. Post Office. Even more than the fresh infusion of investor cash, the hiring of Vail would prove a boon to Bell Telephone and ensure not only its survival but also its growth into a dominant corporation and an industry monopoly. The scrappy little underdog that was Hubbard, Sanders, Watson, and Bell's undersized, underfunded company was about to be radically transformed.

CHAPTER 6

Building a National Network

Theodore N. Vail arrived at Bell Telephone determined to aggressively fight Western Union's patent infringement as well as beat it soundly on the technology front. While Vail took on the telegraph giant in court—with no sense of the intimidation, deference, or fear one might expect from a weak and fledgling start-up—he also threw Bell Telephone's energies into research and development. He was determined to perfect Bell Telephone's equipment before the inventive minds at Western Union—Thomas Edison's, in particular—established the industry standard with a new and superior carbon button transmitter. In the end, rather than start from scratch in the laboratory, Bell Telephone acquired the rights to another inventor's improvement on Edison's unpatented carbon button transmitter.

TROUBLE IN THE FIELD

Vail also sought to raise the courage and spirits of Bell Telephone's field agents, who were trying to establish competing telephone exchanges nationwide. Some of them yielded quickly to Western Union's territorial bullying. Vail sent each agent a copy of Bell's telephone patents as a reminder of exactly which company held rightful claim to the new technology. He sternly insisted to them that, as quoted by Casson, "We have the only original telephone patents, we have organized and introduced the business, and we do not propose to have it taken from us by any corporation."

The field agents had every reason to feel daunted, intimidated, and demoralized, however. Bell's manufacturing partners were not producing the new carbon receivers any faster than they had made the older magneto models. Only 10,000 Bell telephones had been shipped and installed—less than one-fifth of Western Union's total. This shortage of hardware coincided with Western Union's continuing success in staking out territory and creating local telephone exchanges before Bell could get there and attract its own customers. Western Union also had the advantage of being able to draw on many influential existing customers who were accustomed to the company's telegraph service, including the national and local news media, railroads, and hotels. In addition, it had an existing national network of wires and sophisticated research, development, and equipment manufacturing departments and contractors. Bell Telephone was beginning from scratch in all areas—infrastructure, product development, and customer cultivation. Western Union had even begun to acquire publicly traded Bell stock in a ruthless attempt to destroy the upstart competitor by acquiring a controlling share of it.

THE MIGHTY FALL

Partly to outmaneuver this tactic and partly to raise more desperately needed funds, Bell Telephone again reorganized. New England Telephone and Bell Telephone were reunited and rechristened the National Bell Telephone Company, with a new president, the Boston financier Colonel William H. Forbes, at the helm.

At this same time, surprising cracks began to appear in Western Union's monolithic façade. In a stroke of corporate poetic justice, it had become the victim of the same hostile takeover tactics it exerted on Bell. A rival telegraph company owner and financier, Jay Gould, sought to buy out Western Union in order to drive it out of business. Meanwhile, Western Union's own chief patent attorney concluded that the chances of success in challenging Bell's patents were small to nonexistent and declared flatly that Alexander Graham Bell was the original inventor of the telephone. Quite suddenly, given the level of antipathy and fierce competition between the two vastly different companies, the mighty and fantastically wealthy telegraph giant conceded defeat and settled with the tiny and impoverished telephone start-up in November 1879.

In exchange for yielding all of its telephone-related patents and patent claims, Western Union would receive 20 percent of the income generated by customers' rental of Bell equipment throughout the next 17 years of Bell's patent protection. Most importantly, Western Union would sell its burgeoning telephone network—56,000 wired and operational telephones nationwide—to the National Bell Telephone Company. In return, Bell Telephone agreed to stay out of the telegraph business. Even at this crucial point in the history of technology, few seemed to realize that the telegraph business was being edged out by the telephone that Western Union now officially declared had been invented by Alexander Graham Bell, whose ensuing patents it agreed were valid.

Although patent lawsuits not associated with Western Union would continue to pester Bell Telephone until 1888, the removal of its far wealthier and more powerful competitor finally allowed the small company to expand and flourish. Less than two years after the settlement with Western Union, the newly renamed American Bell Telephone Company issued its first annual report to shareholders. It simply but proudly proclaimed the good news: Approximately 140,000 Bell telephones were in use nationwide, nearly all American cities with populations over

10,000 had telephone exchanges, and the company had begun to turn a profit.

LOOKING AHEAD AND MOVING FORWARD

It seemed as if victory could safely be declared and a satisfied sense of achievement enjoyed. Theodore Vail, however, was not at all content to rest. In addition to focusing on building and expanding the still rather patchwork telephone network inherited from Western Union into a true coast-to-coast interconnected communications system, Vail began to worry about the floodgates of competition that would open when Bell's patent protection ran out in 1894. Rather than complacently enjoying its current sole possession of the telephone field, Bell Telephone had to plan for the day when it would not be alone in the effort to wire America. Vail was determined to confront and defeat the competition even before it existed.

Vail used the time remaining on the Bell patents to try to gain such dominance over the telephone industry that the company could easily withstand and beat back any competition that would eventually arise. He intended to do this in three main ways: 1) upgrading the telephone equipment, manufacturing, and infrastructure to improve service, reliability, and sound quality; 2) extending telephone service to as much remaining "virgin territory" (localities not yet wired for telephone service) as possible; and 3) consolidating the quasi-independent local service providers under Bell's control and influence.

This three-prong approach was the foundation of Vail's overarching visionary goal: to create a unified national phone network (composed of Bell-controlled local service providers) based on the advent of strong and reliable long-distance, coast-to-coast service. Vail did not merely imagine a national honeycomb of discrete exchanges, in which each subscriber could communicate only with other members of that exchange. Instead, Vail returned to Alexander Graham Bell's dream that someday soon,

Theodore N. Vail *(above)* became the Bell Telephone Company's savior when he joined the organization as its general manager. Determined and unafraid, he immediately set two goals: destroy Western Union's phone service with copyright lawsuits and develop new technology for future products. His forward thinking helped create the most successful telephone service company in history.

an individual in New York would be able to pick up the telephone and hold a perfectly clear conversation with someone in Los Angeles, as if he were speaking to someone on a point-to-point line strung between two adjacent offices.

EQUIPMENT AND NETWORK UPGRADES

Before this lofty goal could be achieved, many technical problems with the telephone apparatus and infrastructure had to be solved. The apparatus itself was still clunky and unreliable. It had evolved from Watson's earlier one-piece magneto-transmitter/receiver into a three-piece, wall-mounted apparatus.

The telephone apparatus was made up of three boxes. The top box contained an electromagnetic power generator, a hand signal crank, and a bell. The middle box featured the transmitter diaphragm through which a caller spoke. The bottom box held a wet cell battery. To the side of this apparatus hung the receiver, a tubelike piece that was held to the ear. The caller turned the crank to signal an operator at the exchange's central switchboard. The operator plugged into that person's phone line and asked to whom the caller wished to be connected. The operator then called that person, notified him or her of the desired call, and connected the two parties' telephone lines (earlier telephone models did not have bells; a caller who wished to notify an operator of his or her desire to place a call had to bang on the transmitter with a small hammer or other blunt object). When the conversation was done, both parties would hang up, but then the caller had to ring the operator again to notify her that the phone call was complete. The operator would then disconnect the two phone lines from each other.

This model was a great improvement on the earlier one-piece telephone, which required the caller to speak into and listen from the same diaphragm opening. This meant that callers couldn't speak and listen at the same time. Yet, the newer telephone was still a clumsy machine, and the entire process of placing a call and having a conversation was cumbersome, especially because many switchboards became hopelessly wire-tangled and frenetic as customer demand continued to rise. The process was made even more frustrating by the poor sound quality of most phone calls. By the end of the 1880s, city streets were a tangle of telegraph wires, telephone lines, electric streetcar cables, and power lines for the new electric light that Thomas Edison had pioneered. In addition to being unsightly, this dense electrical spiderweb generated electromagnetic fields that caused disruptive and often loud static, hissing, and popping on the phone lines, as did approaching thunderstorms. Telephone poles also proved vulnerable to all sorts of human-made and natural attacks, including car accidents, vandalism, and wind, snow, and ice storms.

The solutions to these problems were fairly simple but still an expensive and time-consuming undertaking. At this point, all telephone circuits were one-wire circuits strung from phone to phone and grounded by the earth itself. Unfortunately, the earth conducted all of the electromagnetic activity going on around it. All of the existing one-wire telephone circuits would be replaced with two-wire versions—known as metallic circuits—that were insulated from the crackling force fields of the electrical surges and currents around them. This meant that Bell's existing network of wires had to be doubled. Bell would eventually bury its wires, placing them safely beyond the disruptive effects of weather, electromagnetic static, and human error. The underground wires would be insulated by crumpled paper and air and wrapped in lead.

The use of metallic circuits greatly improved sound quality across short distances, at least. Across longer distances, however, signal strength weakened dramatically until the transmitted voice was entirely inaudible. This phenomenon is known as *attenuation*, and it was the single greatest stumbling block to reliable and satisfying long-distance phone service. A partial solution was discovered in 1884, when copper wires began to be used instead of the galvanized iron or steel ones that had been

BUILDING AMERICA NOW

THE INCREDIBLE SHRINKING DISPOSABLE PHONE

The first telephones were chunky, clumsy, heavy machines. Made of wood and metal, they were mounted on walls or sat sturdily and squarely on desks, unmoving. Even the lighter and more colorful phones introduced in the mid-twentieth century, though made of plastic, were still solidly constructed, durable equipment that often lasted customers an entire lifetime. These phones were never owned; they were merely leased, with rental fees that were paid every month. Other than features like call forwarding, call waiting, and three-way calling, all of which were introduced in the 1960s and 1970s, the land-based telephones made by Bell Telephone and its competitors at the dawn of the cellular age had not developed dramatically since the introduction of touch-tone dialing in 1963.

Though the first cell phones are now derided for their unwieldy size, resembling the mobile phones used by the military in the Vietnam era, they represented a stunning liberation from the telephone cord that bound people to their homes and offices. Within only a few years, they started to get smaller and smaller. Today, a basic cell phone is considerably smaller than a deck of cards and as thin as a slice of bread. They are also owned, rather than rented, by the telephone subscriber. Cell phones are relatively cheap and are often free or obtained at a steep discount with the signing of a cell phone contract. Indeed, these phones are so small and inexpensive, many people view them as disposable. They are not built to last more than a few years, and many cell phone customers get a new phone every year or so. With more than 180 million cell phone users, it is estimated that more than 500 million cell phones are currently dumped in our nation's landfills, and 125 million more are added each year.

used previously. Copper carried the signal better and longer, resulting in far less attenuation and the establishment of the first successful New York-to-Boston long-distance telephone line (a distance of just less than 300 miles). The line still experienced some pops and hisses, and it could handle only one call at a time. Yet, the technical hurdle to long-distance service was largely cleared. The technology simply needed to be further refined.

Other innovations of the telephone apparatus and infra-structure included the abandonment of phone-housed wet batteries in favor of a single powerful battery housed in central exchange offices. This meant that all phones now received their power from the central exchange center. The capacity and efficiency of switchboards were also increased, and congestion of local exchanges decreased. Finally, Bell Telephone acquired the Western Electric Manufacturing Company, Elisha Gray's former company that had been absorbed by Western Union. Western Electric now did for Bell Telephone what it had done for Western Union—it became the sole supplier of equipment to its parent company. Overnight, Bell Telephone's manufacturing capacity increased exponentially, and the wide gap between demand for phone equipment and supply narrowed greatly.

CLOSING THE REMAINING GAPS

Vail now possessed the equipment and infrastructure innovations required to help him forge a national long-distance telephone network. To achieve his other two goals for a national network—extending Bell's territory and increasing its control over service providers—he decided to form a company-within-a-company dedicated exclusively to the building and operating of long-distance phone lines. In 1885, the American Telephone and Telegraph Company (AT&T) was born.

The company that would come to monopolize American telephone service began as a mere subsidiary of the American Bell Telephone Company. Vail served as its president even as he continued to be Bell Telephone's general manager. Often referred to at the time as simply "the long distance company," AT&T's early mission was not much more complicated than piecing together a national phone network from bits and pieces of existing infrastructure and creating new service wherever gaps existed. Its stated purpose, according to Brooks, was that of "constructing, buying, owning, leasing, or otherwise obtaining, lines of electric telegraph . . . and of equipping, using, operating, or otherwise maintaining the same." In short, AT&T sought nothing less than the complete control, whether through ownership, outsourcing, or leasing, of the nation's entire telephone system—service, infrastructure, equipment, maintenance, and repair.

The way that AT&T opened up new territory and consolidated control over the ever-growing and intertwining telephone network was ingenious, if also somewhat ruthless. AT&T salespeople fanned out across the nation and courted local entrepreneurs, urging them to establish local phone companies and providing them with technical expertise and equipment. For these entrepreneurs to be able to provide service, they would first need a license from Bell granting them the right to get into the phone business. Without a Bell license, the entrepreneurs were violating Bell's patents and would be sued. In exchange for the granting of the license, the local service providers would offer a majority or controlling share of the new company to Bell, allowing the parent company to influence and dictate all major business decisions, maintain quality control, and oversee operations. In addition, Bell leased all equipment to both service providers and telephone customers. Anyone who attempted to gain funding for an independent company would be turned down by the banks, which were being pressured by the well-connected, powerful financiers who owned AT&T stock.

In this way, Bell Telephone would, in effect, come to own and control nearly all of the local service providers, while still being able to technically claim that these were independent franchises and Bell merely licensed the products spawned by its patents. During this period of Bell's patent protection, local providers

In order to expand the company's network and improve service, Bell Telephone strung additional phone lines in existing markets and sought out new territories in need of telephone service. Cities like Seattle (above) installed a web of wires above their streets to meet the rising demand for the telephone, telegraph, public transportation, and electricity.

had no choice but to play by Bell's rules. If someone wanted to get into the phone business, he or she had to do it with Bell's approval, oversight, and input.

PHILOSOPHICAL DIFFERENCES

Just as Vail's vision of establishing a long-distance service within a consolidated nationwide telephone network dominated and controlled by Bell Telephone began to become a practical reality, he suffered a break with the company that delayed the realization of his dream. Vail was a great believer in extending telephone service everywhere within the United States, no matter how small the town or village, and no matter how few people in the locality clamored for it. In addition to wanting to control as much territory as possible before the Bell patents expired, Vail also truly believed in the public service that telephones represented. He felt that everyone should have access to this form of communication, which could be not only life enriching but also potentially life saving, especially in lonely, remote areas of the country. He began to propose extending service to private customers—homes and families and smaller communities—rather than focusing almost exclusively on businesses in America's cities.

Bell Telephone's directors—Boston financiers enmeshed in the business interests of the urban Northeast—disagreed; they viewed this strategy as an unprofitable overextension of the network. Vail saw a simple but profound and tender human need that should be met, whereas his bosses saw only the likelihood of dwindling income.

Vail also objected to the price gouging that Bell Telephone engaged in, given its status as a monopoly. He did not believe that Bell should charge high rates just because it could. It provided an essential service to the public, who had no choice but to use Bell's services; as a result, Vail argued, the company had an obligation to price the service fairly and accept more modest profits. He argued for a per-call fee rather than the existing flat-rate charge. Customers paid a monthly fee for unlimited calls. Most customers did not yet use the telephone enough to get their money's worth. Vail felt that a per-call charge would lower rental fees and encourage households and small businesses to subscribe. It would also make people think before they used the phone, rather than clogging switchboards with idle chatter.

Again the directors disagreed, as did many Bell managers, who felt that providing low-cost commercial and residential service in smaller communities would not be profitable. The focus, they argued, should instead be on high-fee services for big business customers in big cities. A disillusioned and exasperated Vail resigned from both Bell Telephone and AT&T in 1887. Suddenly, Bell Telephone was without its greatest, most insightful protector against the forces of competition on the very eve of their being unleashed.

CHAPTER 7

A Nationwide Competition

Even as AT&T and its parent, Bell Telephone, neglected potential small business and family customers in America's smaller cities and towns, telephone use grew steadily in the years that led up to the expiration of the Bell patents. As Claude S. Fischer points out in *America Calling*, between 1880 and 1893 the number of telephones in use increased more than 400 percent, from 60,000 instruments to 260,000. On the eve of the patent expiration in 1894, one in 250 people possessed a telephone. Most customers at this point were doctors, pharmacists, lawyers, and businessmen in major cities.

Americans increasingly resented the Bell monopoly. The telephone company's customers objected to the high rates, lack of options, and middling service. Its noncustomers were upset at being denied service and being prevented from using local independent providers who could not serve them because of Bell's patent protection. In 1894, this patent protection came to an end, and the doors to all-out telephone competition were thrown wide open.

THE RISE OF THE INDEPENDENTS

Within several years of the opening of a newly competitive telephone market, about 6,000 independent phone companies sprang up, as did numerous telephone equipment manufacturers that supplied the independents with the necessary equipment. Most of these new phone companies were in underserved or completely neglected rural areas in the Midwest and West and sometimes only had a handful of customers. Sometimes a small group of farmers or local merchants would pool their resources to fund, build, and maintain their own tiny telephone network. In *Telephone: The First Hundred Years*, John Brooks quotes a turn-of-the-century Census Bureau report that describes the typical development of such a homemade phone network:

> A group of farmers who lived within a reasonable distance of one another, having come to the conclusion that telephone service was an essential comfort of life . . . would meet together and arrange to establish a telephone system. . . . If the country was wooded, the farmers making up the association agreed to cut and supply the poles and haul them to the places where they were needed. . . . The farmers' boys and the farmhands did the work of setting the poles and putting on the crossarms. . . . The work of stringing the wires and installing the instruments was taken up by the mechanically-minded farmers and their boys, and in a very short time a complete telephone system was in operation. The switchboard was placed in the house of one of the members of the association . . . and the operation of the lines was attended to by the wife and daughters of the farmer in whose home the board was located.

At this time, it is estimated that almost 18,000 private, rural phone lines existed, with 565,000 telephones plugged into them, and almost half a million miles of telephone wire connecting these rural farms, homes, and businesses. Brooks goes on to

TELEPHONE OPERATORS

Telephone operators sat at large switchboards. When a light became illuminated over one of the peg holes or jacks, the operator plugged a cable in, spoke to the caller, determined to whom he or she wished to be connected, found the appropriate jack hole, contacted that person to alert him or her to the impending call, and then connected the two parties. A light would illuminate to signal the end of the call (in the very early days, the caller had to call the operator back at the end of the conversation to inform her that the call had ended).

The first telephone operators were actually teenage boys. It was believed that they would have the energy, dexterity, quicksilver reflexes, and mechanical know-how to connect hundreds of calls an hour on a switchboard composed of a bewildering maze of thousands of cords and jacks. It turned out, however, that they were often impatient, rude, and foulmouthed to callers.

Women, who soon replaced the teenage boys, provided a warmer human voice for the phone company and injected some sex appeal for AT&T's primary customer base—businessmen. In addition to being asked to connect calls, operators were often asked for the time of day, cooking advice, or train schedule information. Some people, especially rural homemakers and lonely men, called simply to make conversation.

With the advent of automated switching and directory assistance, dial telephones, and customer-dialed long distance, most telephone operators were gradually replaced by computerized equipment, though some are still needed to field customer complaints and help connect callers experiencing difficulty. Dial tones were first introduced with direct dialing and automatic switching. The dial tone indicated to the caller that the line was free—something an operator would have handled previously.

wonder "how many lives were saved by calls for medical help or warnings of approaching tornadoes over such lines, and how many times did those straggling wires, strung and hooked up by barely literate amateurs puzzling over electrical manuals, rescue an isolated farm wife from winter stir-craziness?"

As the 1890s wore on and the twentieth century dawned, some independent phone companies began to compete directly with Bell in local markets—even in Bell's Northeast stronghold, including New York and Boston. A town or city might have two or more distinct phone systems. This offered customers choices and competitive phone rates, but it also presented problems. These networks were not connected to each other, so a Bell customer could not call a customer of an independent phone company, even if they lived next to each other or worked in the same office building. Some people subscribed to both services and had two phone lines installed and received two phone directories.

SERVING THE HEARTLAND AND SPREADING WEST

By this time, American Bell and its network of regional and local franchises had enough financial resources, had gained enough influence with bankers and politicians, and had set down deep enough roots to be able to fight off competition in larger markets. Local courts and town councils tended to deny independent phone companies' applications for operating licenses, franchises, and charters or to string or bury competing phone wires. Bell pressured local banks to deny loans and credit to its competitors.

The new competition did, however, force Bell to finally wake up to the needs and demands of vast numbers of Americans who still lived in premodern isolation from major urban centers and even from their far-flung nearest neighbors. By the end of the 1890s, the company had more than tripled its number of customers and approached one million Bell telephones in service. The independents, however, were not far behind. Collectively, they had placed about 600,000 to 700,000 phones in homes and businesses throughout the increasingly connected and chattering nation.

America was wired for phone service, from coast to coast. The so-called "virgin territory" was dwindling, as more and more localities—even the most miniscule—gained access to phone service. In the meantime, as the local Bells began to pay attention to rural America, AT&T continued its efforts to create a national long-distance network. Throughout the 1890s, it opened important long-distance lines: New York–Chicago, Chicago–Nashville, New York–Omaha, and New York–Minneapolis. Long-distance service lines were slowly spreading west, and more and more of the nation's phone customers could place calls from their homes or businesses to other phone customers thousands of miles away.

AT&T TAKES OVER

If American Bell was to continue this simultaneous expansion of local service to rural and small-town America and the extension of long-distance service from coast to coast, it needed more funds—mainly in the form of investment dollars. In 1899, Bell made the fateful decision to move its main operations from its longtime headquarters in Massachusetts to the more freewheeling and business-friendly environment of New York City. Massachusetts laws made it difficult for the company to sell new stock, which was an important way to raise money.

To effect this corporate move, American Bell simply moved its assets to its New York–based subsidiary, AT&T, which—with the stroke of a pen—became the new parent company of Bell. The sale of stocks in the new phone entity allowed AT&T to increase its available funds by more than 300 percent, to $70 million. Within a year, that figure shot up to $120 million, more than twice the combined amount possessed by all of the independent phone companies. AT&T earned $2 million a year more than all of the independents combined, and it only lowered rates

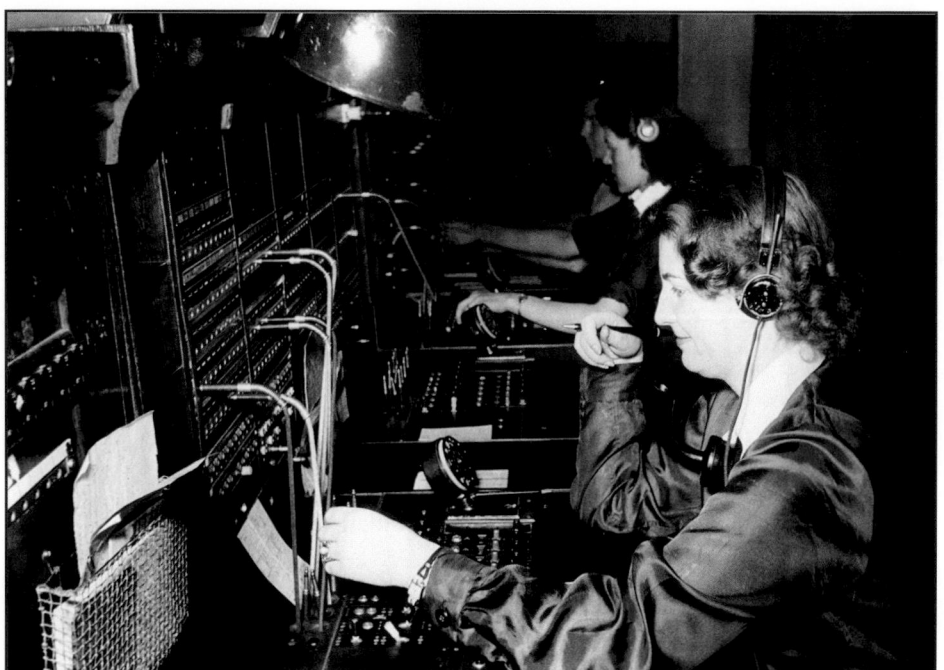

After the use of teenaged boys at the switchboard failed, phone companies turned to women to work the telephone lines *(above)*. Successful candidates had patient dispositions and long arms to reach across the switchboard. These positions were coveted by working women, though most city operators had to handle up to 600 calls an hour while enduring strict rules and unfair pay.

marginally in those areas where it competed directly with other service providers.

As the first decade of the twentieth century wore on, however, the independents began to catch up again. AT&T's expansion efforts were hugely expensive, and the company's debt ballooned. Even as its debt grew, it was forced to lower phone rates to avoid losing customers in an increasingly competitive national environment. Long-distance service was unreliable, time consuming (to get a connection), and prone to poor sound quality. In addition, the West Coast was still not connected to the

East Coast. Though its monopoly was at an end, Americans felt a lingering resentment toward Bell and its aggressive anticompetitive attitude and indifference to its customers. The independents soon had nearly as many phones in service as Bell did—each with about 3 million customers—and investors no longer bought AT&T stock so enthusiastically.

In addition to lowering its rates and showing a new commitment to customer service, AT&T pursued two tactics to ensure its survival and long-term prosperity. More and more communities began to enjoy the choice between two or more phone services. At the turn of the century, more than half of all towns with a population above 4,000 had two or more phone companies that provided local service. AT&T refused to consider allowing its long-distance and local Bell phone networks to be connected to those of competitors. This meant that customers of competing phone companies would continue to be unable to call Bell customers, which was inconvenient to private phone users and a major hindrance to business customers. In addition, any local phone subscriber who wished to place a long-distance call would have to subscribe to AT&T; the independents operated no long-distance lines.

AT&T's second important strategy was the decision that, if it could not beat its competition, it would join them. Rather, it would force the competition to join it. AT&T began to secretly buy independent phone companies, usually through franchises whose connection to AT&T would not be immediately obvious.

THE RETURN OF VAIL

These strategies helped turn the tide in AT&T's favor, but the decisive factor was a brilliant corporate decision made in 1907. The board of directors, now controlled by the financier J.P. Morgan, invited Theodore N. Vail back as president. Vail continued to quietly buy as many independent phone companies as he could. He also bought many of the smaller collective phone exchanges created by farmers or local doctors and businessmen.

At this same time, AT&T's larger competitors began to show cracks. Although their rates tended to be lower than AT&T's, their quality of service was often poor and unreliable. Their profitability was also disappointing and made many of their investors restless and impatient. As investors pulled out, the struggling independents were forced to raise their rates, which angered their customers. Dwindling investment dollars also meant that the independents could not effectively maintain their infrastructure and equipment. For more and more phone customers, there were fewer reasons not to subscribe to AT&T's service.

Vail sought to consolidate this increasing control over the telephone industry by supplying what the independents had begun to neglect and what the Bell network had rarely concerned itself with: high-quality service. Vail poured Bell profits back into research and development, as well as infrastructure maintenance and upgrading of facilities.

In another stroke of ingenious strategy, Vail argued strongly for AT&T's acquisition of Western Union. Morgan was persuaded, and, in 1909, AT&T bought enough of the telegraph company's stock to gain decision-making control. Suddenly, AT&T had access to thousands of miles of additional, already-strung electrical wire and message-transmitting infrastructure, not to mention a massive customer base. AT&T could now offer customers the ability to send and receive telegraphic messages across phone lines, and the shared lines could be used for emergency long-distance telephone or telegraph alerts.

GROWING DOMINANCE

Vail's various strategies worked beautifully. More and more telephone customers switched to the Bell network, either by choice or because their phone service had been bought out by AT&T. More and more, phone customers enjoyed full connectivity within the Bell system and were able to call people who were also served by a local Bell provider.

The development and completion of a coast-to-coast long-distance network was the final element in Vail's plan to establish a universal, nationwide phone network operated solely and exclusively by AT&T. In 1900, an inventor and physics professor named Michael I. Pupin developed the loading coil. These were small electromagnets that could be placed at regular intervals (every 3,000–6,000 feet) along an electric transmission wire. Each coil would supply energy that boosted the transmitted signal and prevented attenuation—the loss of signal strength over long distances. In 1907, another new invention—repeaters—was coupled with loading coils to increase sound quality. Repeaters were vacuum tubes that regenerated weakening signals and amplified sound. The combination of strengthened signals and amplified sound finally made the dream of coast-to-coast long-distance telephone conversations a reality.

AT&T eventually bought the patent rights to the loading coil and repeater, and its own engineers perfected and manufactured working repeaters. The company quickly began to apply the new science to its existing—and often scratchy and static-scarred—long-distance lines. Loading coils and repeaters allowed for the opening of a New York–Denver long-distance line, and Vail publicly promised the establishment of a New York–San Francisco line by the opening of the Panama-Pacific Exposition (celebrating the opening of the Panama Canal), scheduled for 1914.

AT&T now controlled the majority of American phone and telegraph lines. Its call quality and signal strength were increasing dramatically. Its commitment to good service, warm customer relations, and cutting-edge technological development was renewed. Its competitors, as a result, were withering on the vine. Nothing seemed to stand between AT&T and a permanent telephone monopoly. Nothing, that is, but the United States government.

THE KINGSBURY COMMITMENT

AT&T's ruthless acquisition of rival phone companies, its refusal to interconnect with those competitors it could not buy or otherwise drive out of business, and its recent acquisition of Western Union transformed the long-standing low rumble of resentment against the phone giant into rapidly growing and increasingly loud shouts of "monopoly" and calls for trust-busting punishments. The U.S. attorney general concluded that AT&T appeared to have violated antitrust laws, and the Interstate Commerce Commission launched an investigation of the company's competitive—or anticompetitive—practices. In 1912, the federal government sued AT&T for violating antitrust laws.

Rather than risk a costly and ultimately unsuccessful round of lawsuits with the government that would probably result in the breakup of AT&T, Vail instead offered a limited surrender to government control. This surrender would still grant AT&T a telephone monopoly, but a monopoly grounded upon reasonable profit limits, attentiveness to customer needs, and a responsible sense of public service.

Under Vail's direction, AT&T's vice president, Nathan C. Kingsbury, extended a peace offering of sorts to Attorney General James McReynolds in December 1913. In this letter, which came to be known as the Kingsbury Commitment, AT&T offered a three-pronged promise: In exchange for the government dropping its lawsuit, the telephone company agreed to sell its controlling shares in Western Union, cease buying up independent phone companies without prior government approval, and offer its competitors' customers access to its long-distance phone lines. Vail was gambling that this unconditional surrender of AT&T's plan to create a telecommunications monopoly encompassing both national telephone and telegraph service would generate new governmental and consumer goodwill, and AT&T would emerge stronger than ever.

Whether or not this competition was good for AT&T, it certainly was beneficial to customers. Rates in New York City fell about 40 percent between 1905 and 1915, and average residential rates nationwide fell by 50 percent between 1894 and 1909. Not only did non-AT&T customers now have access to long-distance phone lines (AT&T charged its competitors a per-call fee for use of its wires), the new connectivity also allowed AT&T customers to call non-AT&T customers without having to subscribe to two separate phone services.

In its ongoing attempts to dominate as much of the telephone industry as possible, AT&T expanded its network's coverage,

By concentrating on technological development, AT&T was able to establish a long-distance network that fulfilled Alexander Graham Bell's longtime dream of coast-to-coast telephone communication. *Above,* a giant model of a candlestick telephone commemorates the creation of the first transcontinental phone call, at the 1915 Panama-Pacific Exposition in San Francisco.

wiring more and more small-town and rural areas. Public phones began to appear in stores, particularly pharmacies (which needed phones to communicate with doctors who called in prescriptions), and their use was often offered free to store customers. Eventually, however, many of these public phones were converted into pay phones. Urban middle- and lower-class residents tended to use these phones whenever they needed to make a call, or they would rely on the kindness and generosity of a friend or neighbor who had a phone in his or her apartment. Party lines—in which several customers who lived or worked in separate locations could share a single line and its costs—also became popular, despite the occasional frustration of having to wait one's turn to place a call on the shared line and the uncertainty about whether fellow subscribers were eavesdropping on your conversation. The wealthier residents of cities could afford to install phones in their homes and pay a monthly flat fee, and farmers continued to string up their improvised lines and patch them into the Bell system. The telephone was quickly becoming an everyday tool for most Americans, not just a rich person's luxury or a businessman's toy.

AT&T, COAST TO COAST

It could be argued that AT&T's acceptance of a competitive telephone industry actually improved its marketing, manufacturing, and service skills. Indeed, its telephone lines reached more customers than ever, in more places, and the telephone began to enter the fabric of daily life—demand for and delight with the technology was ever increasing. The extent to which AT&T had mastered the telephone was dramatically demonstrated on January 25, 1915, at the opening of the Panama-Pacific Exposition.

As Vail had promised, he had driven AT&T engineers hard to solve the problems of signal attenuation and create the technology and materials necessary to open a New York–San Francisco telephone line. It would be the world's longest telephone line; it would require 2,500 tons of copper wire, somewhere between 130,000 and 180,000 telephone poles, and hundreds of thousands of loading coils placed every eight miles. Three vacuum tube repeaters were placed at thousand-mile intervals to boost signal strength. The new line would connect AT&T's network that had spread outward from New York City toward the West with a separate West Coast network developed by its Bell subsidiary Pacific Telephone. With everything in place, all that remained was to perform a high-profile, attention-getting test.

To boost the carnival-like showmanship of the event and to complete a sentimental full circle, Vail enlisted the aid of two men who knew a little something about both telephones and dramatic salesmanship—Alexander Graham Bell and Thomas Watson. A four-way call was arranged between Vail on Jekyll Island, Georgia; President Woodrow Wilson in the White House; Bell in AT&T's New York offices; and Watson in San Francisco. All four agreed that they could hear one another clearly, as if they were in the same city. At the prompting of a New York reporter, the two old friends and inventors even engaged in a sort of comedic routine, with Bell saying, "Mr. Watson! Come here. I want you!" Without missing a beat, Watson replied, "It would take me a week to get there now!"

The new phone line would not be cheap or accessible to the average phone customer. The first three minutes of a New York-to-California call would cost more than $20 (more than $440 in 2008). Over time, however, the rates would come down. The most important thing was that AT&T had achieved what Vail had set out to do—literally and figuratively unite the country from coast to coast, drawing all Americans together in a humming web of conversation and community. The system was not yet as unified as Vail wanted it to be, nor as exclusively controlled by AT&T as he still hoped it would become. Yet AT&T was the dominant and presiding figure over a new industry that was taking America by storm and quickly becoming an indispensable part of the nation's essential nerve system.

A WIRED NATION AND A "NATURAL MONOPOLY"

America was now wired for conversation, gossip, business transactions, information, and emergency calls. Its most remote areas would never again feel like the isolated, lonesome, disconnected places they had once been.

The opening of the New York–San Francisco line suddenly made the telephone a glittering, exhilarating national phenomenon, rather than a humble, unassuming—and often frustrating—instrument for local communication. Suddenly, the

TELEPHONE POLES AND LINEMEN

America would not have become wired without two unsung heroes of telephone history—the telephone pole and telephone linemen. Poles were first used to string telegraph wires. Stringing them high above the ground protected them from human traffic and accidents on the ground and prevented short-circuiting. With the introduction of the telephone, copper wires were strung on telephone poles that ran through city streets and across the vast countryside.

The poles in the city, where the bulk of early telephone subscribers worked and lived, became a major nuisance. To accommodate the growing number of phone lines, the poles rose as tall as 90 feet and included 30 crossarms strung with 300 or more wires. The web of wires was so dense that the sky could actually be blocked from view. Northeastern wind, ice, and snowstorms frequently caused downed or frozen wires, which interrupted service across wide areas for extended periods of time. In 1885, New York State passed a law intended to force the phone companies to bury their wires, but the law was repealed before it had any effect. In 1890, however, Bell Telephone did begin to place some of its phone lines underground.

In many areas of the country, however, telephone poles persist to this day, especially in rural areas and older suburbs. Most are telephone seemed to embody the semimythical American dream: visions of open spaces, travel, mobility, exploration, expansion, and footloose fortune-seeking. It allowed people to conduct business and maintain close relationships across vast distances, to bridge thousands of miles instantaneously and from the comfort of home. With the advent of coast-to-coast long-distance phone lines, AT&T was, in most Americans' minds, no longer just a telephone company. It was *the* telephone company.

The Kingsbury Commitment established a relationship between the AT&T/Bell system and the government that would

still wooden—made from pine, fir, or cedar trees—and are treated with creosote to combat damage from woodpeckers, insects, fungi, and fires. They are now strung with coaxial cable or fiber-optic cable, rather than with copper wire. Newer poles may be made of steel, concrete, or fiberglass and strung not only with phone lines but also with cables for television, streetlights, traffic lights, and electricity. There are now more than 160 million telephone/utility poles in the United States.

Telephone linemen are the men and women responsible for installing and repairing aerial (strung on poles) and underground phone lines, as well as the various equipment associated with the phone lines, such as conduits and insulators. They also maintain the poles themselves. They must respond to problems and emergencies that arise anywhere along the hundreds of thousands of miles of telephone cable strung and buried throughout the nation and even across oceans. It is a dangerous job that exposes linemen to all kinds of weather, dangerous levels of electricity, and the very real possibility of lightning strikes and falls from a great height. As many as 50 linemen a year died on the job in the early years of telephone service.

endure for the next seven decades. AT&T was given free reign to absorb competing companies if it wished and to essentially oversee and direct the remaining noncompeting independents patched into the Bell system. The government had concluded that—as long as prices remained fair and reasonable, service was extended to all who wanted it, and quality was high—the American people would benefit most from a unified, consistently operated, and rationally laid-out national telephone network, rather than a chaotic patchwork of service providers, varying phone rates, and territorial turf battles and confusion.

As a result, in exchange for granting some government regulation—particularly of its pricing of services—AT&T was granted a near monopoly over the telephone industry. Some competition was permitted, mainly in areas that AT&T/Bell was not interested in servicing (usually remote, unprofitable regions). In addition, AT&T was again permitted to buy controlling shares in many independent phone companies. The system that resulted was the creation of a network of regional Bell companies—with names like New York Telephone, New England Telephone and Telegraph, Mountain States Telephone and Telegraph, Southwestern Bell Telephone, and Pacific Northwest Bell Telephone— that were at least partly owned by AT&T.

These regional Bell companies provided local phone service and paid AT&T a percentage of their revenues as a licensing fee, as well as a portion of their stock dividends. In exchange, AT&T provided the local Bells with all of the necessary telephone expertise, the benefits of its research and development, and equipment manufactured and supplied by Western Electric—wires, cables, circuits, switchboards, poles, and phones that would in turn be leased to Bell customers. AT&T was also obligated to provide the local Bells with access to its long-distance lines and circuits, so their customers could make interstate calls to anyone anywhere in the United States (and, later, anywhere in the world).

With the exception of a single year during World War I when the government took over phone operations (and during which time rates went up and profitability and efficiency fell), AT&T enjoyed this monopoly through the 1970s, despite several serious court challenges on antitrust grounds. By the interwar years, 80 percent of the phones in use were Bell telephones, and nearly all of the non-Bell phones were connected to Bell lines. Competition had led to the widespread, coast-to-coast wiring of America. Whatever telephone lines AT&T had not strung itself, it acquired through decades of absorbing the assets and infrastructure of independent companies, beginning with Western Union in 1879. It now owned or controlled nearly all of the tens of thousands of miles of telephone wire strung from sea to sea.

CHAPTER 8

An Age of Innovation

The nation was wired. The infrastructure was in place and regularly maintained by a single company. All that remained for AT&T was to expand the national network—and its profits—by increasing demand for its equipment and services.

This became particularly urgent during the lean years of the Great Depression, when national spending decreased dramatically and consumers simply did without many goods and services. Businessmen were told how the telephone would boost their efficiency and productivity, decreasing expenses on things like stationary and stamps and eliminating communication lag time. Person-to-person communication would win new clients, they were assured, and the absence of a telephone would put one at a disadvantage with competitors who did use phone service. Private households were pitched telephone service in more emotional and sentimental terms, with a focus on maintaining far-flung friendships and family relations, fulfilling local social obligations, providing an essential labor-saving device to busy homemakers (who could order groceries and schedule repairmen over the phone), and serving as a vital lifeline in case of emergencies.

The ad campaigns seemed to work. Because the phone was born there, and it was a heavily urbanized and commercial region, the Northeast had long enjoyed comprehensive telephone service. The West and Midwest soon began to catch up, however, and then outpace the Northeast in terms of new telephone customers, especially following World War II, when postwar affluence and suburbanization made telephone service more affordable and essential. Indeed, in 1946, 25 Bell telephones were installed every minute during work hours. The South would lag far behind well into the twentieth century, due to the fact that it was largely rural, nonindustrial, and generally poorer than the rest of the nation until the second half of the century, when the arrival of the Interstate Highway System brought more investment, companies, jobs, and middle- to upper-class homeowners to the region.

INTERNATIONAL CALLING

With the telephone infrastructure in place and the nation largely wired for service, AT&T turned its attention for much of the rest of the twentieth century to attracting as many new customers as possible and patching them into the Bell network. It also sought to maintain and upgrade the network, through the everyday repair and maintenance work of a nationwide corps of electrical engineers, technicians, and linemen. It also rolled out technological innovations that made the network more efficient, high capacity, and extensive; service faster and more reliable; and sound quality far clearer.

These innovations began with a bang in the Roaring Twenties. Harnessing the emerging technology of radio—which was developed in part by AT&T and built on Alexander Graham Bell's pioneering work in telephony—AT&T began to provide transatlantic telephone service to Great Britain in 1927, transmitting the signals via two-way radio. This service was later expanded to include other British Commonwealth countries, including Canada, Australia, Kenya, Egypt, and South Africa. AT&T established a U.S.–Paris service in 1928. Telephone service to Japan was achieved in 1934, and the first round-the-world call was placed in 1935. Two AT&T officials in adjoining New York City offices spoke to each other on the phone, but each caller's voice traveled around the world in a massive, instantaneous loop before it was transmitted into the listener's receiver.

ACHIEVEMENT IN DARK TIMES

Also in the 1920s, in response to growing demand for service and increasing telephone traffic, AT&T established its first Traffic Control Bureaus in Chicago, Illinois; Cleveland, Ohio; and New York City. These offices were tasked with managing the national long-distance network and the interconnections of all the regional Bells. They were nerve centers, in constant contact with one another. They reported on call traffic patterns, system failures, equipment malfunctions, weather-related damage or disruption to the system, and downed wires or poles. They could also contact switching centers around the country to request that they handle rerouted calls to lighten the load on an overburdened switching center somewhere else in the system. Like air traffic controllers or railroad system managers, the AT&T Traffic Control Bureaus ensured that calls kept moving, bottlenecks were eased, and malfunctions or structural problems in the system were fixed promptly, with minimal disruption or inconvenience to customers.

In 1927, AT&T staged the first public demonstration of television, transmitting via cable the live image of Secretary of Commerce Herbert Hoover speaking on a telephone. A Bell Lab scientist named Herbert Ives had developed the science of image transmission by combining the photoelectric cell with a vacuum-tube repeater to transmit images over telephone wires. The first application of this technology was to quickly send still pictures to news agencies, but Ives and Bell Labs soon began to work on sending moving images across telephone wires. In 1929, AT&T presented the first public demonstration of color television. The Depression stalled the development of commercial television, however, and other companies pursued and perfected the technology in the years that followed. Beginning in 1948, AT&T did, however, offer networking services to television networks, which allowed them to transmit their programming to local affiliates from coast to coast.

During the years of the Great Depression and World War II, AT&T suffered sharply decreased demand for phone service, wartime rationing and shortages of essential telephone network materials, and manpower loss. Yet the company still managed to introduce new technology and upgrade its network infrastructure. Beginning in the mid-1930s, AT&T began to experiment with coaxial cable, which was the world's first broadband transmission medium.

The network's phone lines were, at this time, made of copper and could carry several calls on a single wire (previous to this, only one call at a time could be carried on the wire). Some lines were buried to protect them from the weather and to allow more wires to be strung in less space, but most telephone wires were still strung on poles. Coaxial cables included two conductors wrapped in layers of insulation. Their capacity was far greater than that of copper wires, and they were significantly less thin. These cables, which were buried in the ground, could initially carry almost 500 conversations at once, plus one television broadcast. Eventually, the technology improved to the point that coaxial cables could handle more than 130,000 calls at once and thousands of television broadcasts. As much as AT&T is celebrated for wiring America for phone service, it also played an indispensable and unsung role in wiring America for the emerging medium of television, a new technology that would alter the American cultural landscape every bit as much as the telephone had.

MICROWAVES AND TRANSISTORS

AT&T emerged from the Depression and World War II weakened but encouraged by Americans' desire to again spend money and enjoy the good life following such a prolonged period of deprivation, gloom, and sacrifice. It also came away from the war with a handful of exciting new technologies that its research and development arm, Bell Laboratories, had developed for military use. The very first commercial mobile phone was introduced in 1946, for use in cars and linked to the Bell network by radio signals. Microwave relay also came into use following World War II. Phone calls and television broadcasts could be transmitted via radio signal through a series of radio towers, which were cheaper and easier to build and maintain than the massive excavation project that was required to bury hundreds and thousands of miles of coaxial cable (often in rugged terrain, remote areas, or inaccessible landscapes) and ensure their continued proper functioning.

By the 1970s, microwave relay handled 70 percent of AT&T's phone calls and 95 percent of television transmission (it would eventually be supplanted by fiber-optic cables and satellites). A New York-to-Boston microwave relay system was established in 1947, and by 1951, the first transcontinental microwave relay was built. This relay included 107 relay stations placed at 30-mile intervals; it could handle 600 phone calls and two television broadcasts. By the end of the 1950s, more than 400 microwave relay stations operating around the country accounted for 25 percent

---

(opposite) Although the Great Depression resulted in consumers spending less on goods and services, AT&T was able to expand its business during this era with effective advertising campaigns. By explaining the practical uses of the telephone for commercial and personal purposes, the telephone giant established a national network that wired the United States.

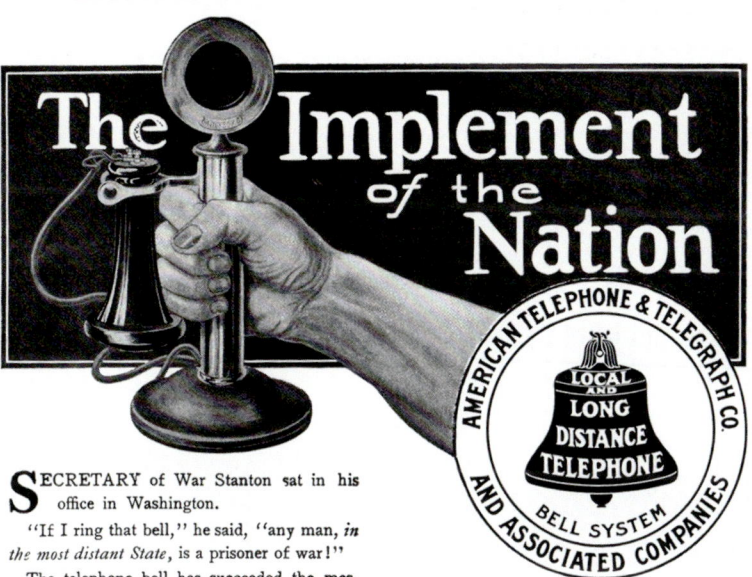

The Implement of the Nation

SECRETARY of War Stanton sat in his office in Washington.

"If I ring that bell," he said, "any man, *in the most distant State*, is a prisoner of war!"

The telephone bell has succeeded the messenger bell.

Business has succeeded war.

If *any man* in the Union rings the bell of his Bell Telephone at his desk, any other man *at the most distant point* is at his instant command.

That is the Bell Companies' ideal—that you may take the receiver off the hook and get into communication with *any man*, even in the most distant State.

That is the really *universal telephone* that the Bell Companies set as their goal at the beginning. It is so far realized that already 20,000,000 voices are at the other end of the line, all reached by the one Bell system.

The *increased efficiency* of the individual, of the lawyer or bank president or corporation official; the increased efficiency of the *nation as a whole*, because of the development of the Bell system, can hardly be estimated.

It certainly *cannot* be overestimated.

The president of a corporation to-day could not be the president of such a corporation without it.

The modern corporation *itself* could not exist without telephone service of national scope.

Corporation officials could not have transacted business quickly enough by old methods to reach the totals which alone are accountable for our remarkable commercial development as a nation.

The wheels of commerce have been kept at the necessary speed to provide this swift development by the universal telephone.

The mere item of *time actually saved* by those who use the telephone means an *immense increase* in the production of the nation's wealth every working day in the year.

Without counting the convenience, without counting this wonderful increased efficiency, but just counting *the time alone*, over *$3,000,000* a day is saved by the users of the telephone!

Which means *adding $3,000,000 a day to the nation's wealth!*

The exchange connections of the associated Bell Companies are about 18,000,000 a day—the toll connections half a million more. Half of the connections are on business matters that must have prompt action—either a messenger or a personal visit.

Figured on the most conservative basis, the money value of the *time saved* is not less than ten cents on every exchange connection and three dollars on every toll, or long distance connection—figures that experience has shown to be extremely low.

The saving *in time only* is thus $1,800,000 daily on exchange messages and $1,500,000 on long distance messages—this much added to the nation's productiveness by the Implement of the Nation, the Bell Telephone.

American Telephone & Telegraph Company

of AT&T's long-distance lines. In the 1960s, AT&T provided the microwave transmission system with a cable backup in the event of nuclear war. Bell Labs developed an atomic blast-resistant coaxial cable. AT&T laid 2,500 reels of the cable in a 4,000-mile-long trench. It even built 11 manned test centers along the line, each one 50 feet underground and equipped with living quarters, food, water, and filtered air.

Perhaps the greatest Bell Labs' triumph ever, however—one that would lead directly to our own high-tech, digital, computerized age—occurred in 1947, when a team of Bell researchers perfected a working transistor, which is a small but powerful semiconductor. Their transistor was the world's first solid-state amplifier (which allowed for far greater amplification of a signal than vacuum tubes could achieve) and electronic switch. Analog circuits use transistors as amplifiers, and digital circuits use them as electronic switches. Because transistors were so small, they made possible the integration of many circuits within a space the size of the top of a pencil eraser, which in turn allowed for miniaturized electronic devices such as hearing aids, handheld radios, silicon chips, and electronic telephone switches. It also led to the improvement in sound and performance of stereos, radios, and televisions. Eventually, the design and operation of satellites and rockets would be made possible by transistors. Transistors are to computers as the cell is to the human body—they are generally referred to as the basic building block of computers.

END OF AN ERA

The dawning of the electronic age spelled the beginning of the end for one of AT&T's most iconic and beloved analog features— the operator who placed every call. By the early 1940s, more than half of America's telephones were rotary dial. People no longer had to speak to an operator to place local calls, though the operator still connected those calls. By the early 1950s, more than 80 percent of phones in American households were rotary dial, and almost 90 percent of them could place local calls directly, without operator assistance. Most callers still needed to speak with an operator to place long-distance calls.

This situation changed with the introduction of direct-distance dialing (DDD) in the 1950s, which allowed AT&T customers to simply pick up the phone, dial the desired number, and speak to someone thousands of miles away, without waiting as long as several minutes to reach an operator and wait for him or her to make the connection (although international long-distance

TELEPHONE NUMBERS

As a result of the advent of dial phones in the 1920s—which allowed customers to place their own local calls without operator assistance—phone numbers needed to be created. Operators previously connected one caller to another, but now each telephone user would be assigned a number that other people could dial to reach him or her.

The first telephone numbers were combinations of letters and numbers. Two or three letters would indicate what exchange the customer was in, and four or five digits would be the customer's unique number. The exchange grouped customers in a geographical region, and its two-letter code often referred to some aspect of that region. So, for example, a resident of the Columbus Circle area in New York City might have the phone number CI9–0463 (CI stood for "circle"; the actual telephone number was 249–0463). A resident of the city's Murray Hill section might have the number MU3–8829 (683–8829). When people gave out their phone numbers, they'd say, for example, "I'm at Murray Hill 3–8829" or "Niagara 3–5834" (NI3–5834, or 643–5834). The purpose of this system was to help callers remember phone numbers. Place names and four or five digits, it was believed, would be far easier to remember than a string of seven digits.

Beginning in 1958, the system began to switch over to all-number telephone exchanges. By the mid-1980s, the entire North American phone network had been converted to all-number dialing.

calls were placed through operators until 1970). At this point, operators still connected the phone call, but they did not need to get on the line or speak to the customers. In the 1960s, however, automated, computerized call switching, which utilized the transistors developed by Bell Labs, put thousands of operators out of work. The last public operator-staffed switchboard was shut down in 1978.

FROM THE SEA TO THE STARS

As these advances in telephony were introduced and embraced enthusiastically by appreciative customers, the transatlantic two-way radio telephone line established in 1927 began to seem increasingly arcane and problematic. The service was unreliable. It was noisy and subject to atmospheric interference. It had relatively low caller capacity and poor signal quality. The problem was figuring out a way to create a transatlantic cable that would maintain signal strength across thousands of miles and require very little maintenance. Locating and servicing a problem on a cable that rested on the ocean floor several miles below the water's surface would be a logistical financial nightmare and would disrupt all transatlantic service for a prolonged period.

The engineers at Bell Labs eventually solved both problems with the creation of a single device—an underwater repeater that could amplify transmitted signals and function properly for at least a 20-year period. One of these torpedo-shaped, 3-foot-long, 500-pound repeaters would be placed along every 10 to 40 miles of underwater cable. Each one cost $100,000. AT&T would soon recoup this investment, however. Within three months of the transatlantic cable's opening for public use in 1956, it handled 870 calls a day; it had a capacity of 36 calls at one time, at a price of $12.00 for the first three minutes. A transpacific line that linked Japan to Hawaii and Hawaii to California followed in 1964.

From the depths of the ocean, AT&T next set its sights on space. Beginning in the 1950s, Bell Labs began work on prototypes of communications satellites that could be launched into orbit around Earth and serve as relay stations for microwave radiotelephone signals. The signal from a call placed in New York could be bounced off of a series of communications satellites and ground stations and then received on the opposite side of the globe. This would be far quicker and cheaper than relying on transatlantic cables or thousands of relay towers. It would also relieve the traffic congestion on AT&T/Bell's telephone cables and radio circuits, which, because of steady customer growth and telephone usage, would soon be operating at full capacity.

In August 1960, with the help of the National Aeronautics and Space Administration (NASA), AT&T launched *Echo I*, a passive communications satellite that could reflect telephone signals back to Earth but could not actually receive and retransmit them. The signal would weaken as it traveled from ground station to satellite and back to ground station because it was not being retransmitted. It did not look like the satellites we are familiar with today; rather, *Echo I* was essentially a huge, plastic-coated balloon.

Two years later, AT&T and NASA launched *Telstar I*, an active satellite that could receive and retransmit signals (whether they be radio, telephone, or television signals), thereby maintaining signal strength from ground station to satellite and back to another ground station. Its electrical system was composed of thousands of diodes and transistors, and it was powered by 3,600 solar cells. A telephone call placed from an AT&T ground station in Andover, Maine, was beamed to *Telstar* and retransmitted to Washington, D.C., where Vice President Lyndon B. Johnson fielded the call that came through loud and clear. In addition to improved long-distance and international calling, communications satellites would soon allow for live worldwide broadcasts of events that occurred halfway around the world, such as breaking international news, Olympic Games, a 1973 Elvis Presley concert in Hawaii, the 1981 wedding of Prince Charles and Lady Diana Spencer, and the simultaneous broadcasts of the concurrent Philadelphia and London Live Aid concerts of 1985.

The bright promise of the telecommunications future was further heralded by AT&T's introduction of the first commercial modem at the same time it was developing communications satellites. The modem allowed two or more computers to

BUILDING AMERICA NOW

TELEPHONE VERSATILITY

Before the cell phone age began, land-based telephones primarily carried voice data—the sound of the human voice speaking to another human. Yet, even within this seeming limitation, many innovative services were introduced. One could call a certain phone number to get the exact time of day. The Weather Bureau provided updated local weather forecasts by phone. "Dial-a . . ." services, such as "Dial-a-Prayer" and "Dial-a-Joke," were introduced. Innumerable crisis hotlines and medical information services were established. The phone was even used by local businesses to advertise their services to unsuspecting phone customers.

Yet, the bewildering array of features that today's cell phones offer and the number of services one can access through them is an exponential leap forward. In addition to providing basic telephone service, cell phones also double as digital cameras, music players, movie players, camcorders, address books, instant text messengers, yellow pages, Internet portals, and Web browsers. Some phones, such as Apple Inc.'s wildly popular iPhone, offer all features in one compact, stylish package. It should come as no surprise that AT&T, always a savvy predictor of market trends, ensured that it would gain a piece of this burgeoning market. It was the first company to offer mobile phone service, way back in 1946 (for use in cars), and the first to open a commercial cellular telephone service, in 1983. It is also the exclusive distributor and carrier of the iPhone.

communicate with each other via telephone wires (a decade later, in 1971, Bell Labs developed the computer operating system UNIX, which would become the most widely used operating system on the Internet and many personal computers, assist in the creation of Internet networking, and lead to Bell Labs' development of the C programming language). Initially, modems were useful only to businesses with large mainframe computers. With the eventual introduction of the personal computer, however, the modem helped usher in the Internet age that would revolutionize the telecommunications industry and profoundly alter global culture.

NEW PRODUCTS AND SERVICES

Even as AT&T was reaching for the stars and spinning the first threads of the World Wide Web, however, it also remained mindful of its customers back on Earth and their more mundane interests. Though AT&T was a monopoly, the company realized it still needed to court its customers and offer inviting products and services. In the late 1950s, AT&T began to offer phones in colors other than the traditional black, introducing colors like white, beige, pink, turquoise, and blue. In 1963, it offered touch-tone phones (using numbered buttons), and the old, clumsy rotary dial phones gradually began to disappear. In the mid-1960s, the network's first computerized Electronic Switching System was activated, which improved call routing and traffic management, and eventually allowed for such services as speed dialing, conference or three-way calls, and call forwarding (digital electronic switches eventually allowed 350,000 calls an hour to be connected). At this time, nearly all of AT&T's customers enjoyed direct dial local calling, and the vast majority—80 percent—could also place long-distance calls directly, without operator assistance.

This customer commitment and innovative product and service development resulted in the company's continued dominance of the telephone industry. In any area in which it offered service in the 1960s, AT&T/Bell enjoyed the patronage of at least 85 percent of households (20 years earlier, it could claim only 50 percent of households in its service areas). It controlled more than 90 percent of all telephone-related products, equipment, and manufacturing (produced by Western Electric, including phones, switches, wires, cables, etc.); 98 percent of all long-distance lines; and all of the transoceanic radio telephone systems. AT&T/Bell collected 90 cents of every dollar made in the telephone business in the United States.

This was the high-water mark for AT&T and the Bell system. It owned and controlled the nation's telephone infrastructure. The vast majority of Americans subscribed to its local and long-distance services. Its products were popular, reliable, and innovative. It was on the cutting edge of emerging telecommunications frontiers, including satellite and computer technologies. Yet, beginning in the late 1960s, the telephone giant suffered a series of attacks and setbacks that radically diminished its industry dominance and led to renewed wide-open competition— at the very moment when new technologies made traditional landline-based telephony merely one of many forms of high-speed communication.

CHAPTER 9

The Industry Transformed

In 1969, serious service problems began to arise throughout the telephone network, including lack of dial tones, crossed signals, overloaded circuits, calls that failed to be connected, false busy signals and false rings, line noise, dead lines, and inaccurate billing. The culprit seemed to be some combination of maxed-out system capacity and an infrastructure that suffered from neglect and poor maintenance. In 1971, perhaps in response to these widespread and enduring service breakdowns, the government granted the three-year-old company Microwave Communications Incorporated (MCI) permission to compete with AT&T in the long-distance market.

RENEWED CRITICISM AND COMPETITION

Three years later, the Department of Justice filed an antitrust suit against AT&T. This was partly in response to AT&T's growing movement from pure telephony into electronics, satellite communications, and computer/Internet technology. Yet, it also was in response to a growing scandal surrounding Southwestern Bell and Southern Bell, in which two dismissed and disgruntled company officials admitted to paying bribes to politicians in exchange for legislation and regulatory oversight favorable to the Bells, illegally wiretapping customer phone lines, and maintaining artificially high telephone rates. One of these officials made the admission in a suicide note written before he killed himself by sitting in his running car within a closed garage. The government's antitrust lawsuit dragged on for almost 10 years.

In the meantime, more competitors were allowed to jump into the long-distance fray; one of the most successful was a company called Sprint. A subsidiary of the Southern Pacific Railroad, Sprint could trace its origins to a tiny landline strung in Abilene, Kansas, in 1899, known as the Brown Telephone Company. Sprint began as a puny competitor to AT&T, but it emerged in the late 1970s as a powerful threat, primarily in the major American cities. AT&T was forced to allow MCI and Sprint connectivity to its Bell system, making it possible for local Bell users to bypass AT&T's long-distance service in favor of a competitor without getting an additional phone line. The government also allowed telephone manufacturers independent of AT&T—in particular, General Telephone and Electronics (GTE)—to make and sell phone equipment to AT&T/Bell customers that would be compatible with the Bell network. Soon, customers could buy non-Bell telephones, simply plug them into a jack, and place calls as usual. No longer were they forced to pay a monthly fee to lease a Bell phone that they would never own, no matter how long they leased it.

AT&T had reached the pinnacle of an industry it dominated. It was a giant; however, as in its early days of monopoly under Theodore Vail and J.P. Morgan, the company was increasingly resented—even despised. Politicians and customers alike began to view its legal, government-sanctioned telecommunications monopoly with suspicion and outright hostility. AT&T was losing its grip on the nation. The sentimental and affectionate image of "Ma Bell" created by a grateful public—thankful to be connected to one another in conversation—began to yield to darker imaginings of an evil empire.

A DAZZLING BREAKTHROUGH AND A STUNNING BREAKUP

One of the few bright spots for AT&T and the Bell system throughout the turbulent 1970s was Bell Labs' developmental work on fiber-optic cables. These cables, through which pulses of light were conveyed via ultrapure fibers of glass thinner than strands of human hair, could carry digital transmissions. The wires' capacity was far greater than the analog equivalents of microwave relay and coaxial cable. This meant that digitized data and video could be carried along the wire in addition to the sound of the human voice. Given the emergence of computer and Internet technology and the introduction of personal computers and modems, fiber-optic systems were a monumental breakthrough. If AT&T could ward off the growing antitrust and free-market competitive forces and hold on to its monopoly, it could be in a position to control crucial parts of Internet infrastructure and dominate the growing Internet technology field, just as it had done at the turn of the twentieth century, telephone pole by telephone pole, mile after mile of strung wire.

Unfortunately for AT&T, it was not to be. In 1982, just before it established its first fiber-optic route between New York City and Washington, D.C., the lawsuit with the government was settled. AT&T would survive in name and in fact, but it would be a radically reorganized and diminished entity. It was forced to give up the 22 regional Bells, which would now consolidate into seven independent telephone companies that provided local and regional service. AT&T was allowed to keep its long-distance service, but the doors were thrown wide open to competition. It was also allowed to retain its manufacturing arm, Western Electric, and its research and development branch, Bell Labs. On the day the agreement took effect—January 1, 1984—AT&T's assets fell from $150 billion to $34 billion, a devaluation of more than

400 percent, thanks to the loss of the regional Bells. As Steven Coll says in *The Deal of the Century*: "AT&T . . . was on its knees at last."

Long-distance competition became fierce in the 1980s. Rates dropped as MCI, Sprint, and AT&T fought to attract and maintain

BELL LABS

Bell Labs was the research and development arm of AT&T from 1925 until 1996, when the division, by then known as Lucent Technologies, spun off and became an independent company. The company became part of Alcatel-Lucent, a global communications solutions provider, and the division now goes by the name of Bell Labs again. Following are just a few of the revolutionary, cutting-edge technological innovations it has developed over the years:

★ The first synchronous-sound motion-picture system
★ Vacuum tubes (which made long-distance telephony possible)
★ Negative-feedback amplifier (which made possible transcontinental multichannel telephone transmission, high-fidelity recording, and computer control systems)
★ Two-way radio telephone communications
★ Transmission of still and moving images across phone lines and cables
★ Black-and-white and color television
★ Microwave radio relay
★ The electrical-relay digital computer
★ Computer networking
★ Transistors (crucial to the development of computers, portable radios and stereos, and manned space flight)
★ Coaxial cables
★ Touch-tone telephone
★ Sonar

customers. AT&T saw much of its exclusive territory erode as MCI began to offer international calling in 1983. Sprint beat AT&T at its own game: It established an all-fiber-optic phone network in 1987, two years before AT&T announced its intention to eventually replace its remaining microwave and coaxial

- ★ Lasers
- ★ Communications satellites
- ★ Direct long-distance customer dialing
- ★ Information theory (helped determine the maximum data-carrying capacity for any given communications system)
- ★ Digital transmission and switching
- ★ Charge coupled device (CCD; a solid-state chip that transformed light into electrical information and led to high-definition TV, digital cameras, camcorders, video conferencing, and satellite surveillance)
- ★ UNIX operating system and the C programming language (Made large-scale networking of computers possible, as well as the Internet itself. UNIX is the operating system used by most large Internet servers and business and university systems. C is the world's most widely used programming language.)
- ★ Digital signal processor (found in personal computers, modems, wireless phones, answering machines, voice mail, video games, DVD players, and digital cameras)
- ★ Fax machines
- ★ Modems (which allow computers to communicate with each other)
- ★ Cell phone technology
- ★ Silicone solar batteries (solar cells)
- ★ Light-emitting diodes (LEDs)
- ★ Fiber-optic cables

systems with fiber optics. In 1989, in a largely ignored echo of its glory days, AT&T laid a 3,148-mile-long fiber-optic cable across the Atlantic Ocean, which allowed 40,000 calls at once between the United States and Europe. Capacity would eventually reach 1,000,000 calls at once.

THE ARRIVAL OF THE INFORMATION AGE

Lost in all this swirling chaos and disorienting change is the fact that AT&T had not entirely lost its visionary instincts for identifying and developing "the next big thing." In 1983, in most respects a bad year for the fallen company, AT&T did establish the first cellular phone system in the United States. Cellular phone technology itself was a concept first set forth and tested (unsuccessfully) by Bell Labs, as far back as 1947.

Cell phone signals are basically radio signals transmitted by an extensive network of radio towers. Territory is divided up into "cells," roughly hexagonal areas of about 10 square miles. Within each of these cells a tower and base station are placed, containing radio transmitting and receiving equipment. Fiber optics, cellular phones, and Internet services—all cutting-edge, Information-era technologies that would not have been possible without Alexander Graham Bell's pioneering work in telephony or Bell Labs' subsequent innovations—would preserve AT&T's honored place in the expanding telecommunication universe and marketplace.

By the dawn of the twenty-first century, when more data was being transmitted over AT&T wires than voice communication—and many people worldwide were foregoing landlines altogether in favor of cell phones, Internet-based phone service, and other forms of wireless and digital communication—AT&T had found its way back and reestablished a leading position in the dizzying world of telecommunications. The cellular market alone was huge and still growing; by 2004, 60 percent of Americans owned cell phones. Yet, AT&T now had no shortage of competitors in supplying both landline and cellular service and in the manufacturing of telephones. Sprint and Verizon (formed by the merger of GTE and Bell Atlantic, which was itself an independent, post-AT&T-divestment entity composed of the former Pennsylvania and New Jersey Bells) offered stiff competition in the local

BUILDING AMERICA NOW

CELL PHONE TOWERS

The hundreds of millions of wooden poles and metal microwave radio towers driven into the earth to wire America for telephone service are now being joined by a separate but related network of imposing structures—cell phone towers—built to relay endless streams of data. These steel towers are hundreds of feet tall and are topped with antennas. Several cell phone service providers may share a single tower. The antennas are connected by cables to radio transmitters and receivers and to computerized switching equipment sheltered in box houses on the ground below. When a person makes a call on his or her cell phone, a signal is sent from the cell phone's antenna to the antenna of the base station. The base station then assigns the signal an available radio frequency channel. Transmission and reception of these radio signals transfer the voice data—the words the person speaks—to the base station. The voice data is then sent to a switching center, which transfers the call to its destination—the telephone number dialed.

There are currently more than 175,000 cell phone towers in the United States—more than the number of telephone poles that were required to wire the New York-to-San Francisco long-distance telephone line in 1915. Just like the telephone poles before them, cell phone towers have generated controversy for the way they mar the visual beauty of landscapes and skylines. In response, the cell phone industry has begun to disguise antenna towers as massive pine trees, cacti, church steeples, highway signs, and boulders.

As cell phone towers began to spring up all over the country, AT&T was transformed into a more modern company to ensure a brighter future for Alexander Graham Bell's historic invention. *Above*, a modern cell phone tower, intended to blend in with its surroundings.

and long-distance markets as well as the cellular markets. Other leading cell phone service providers were T-Mobile, Alltel, Nextel, and Cingular (until it was acquired by AT&T in 2007). Some of the leading telephone manufacturers included Sony, Panasonic, GE, and RCA, as well as AT&T. Leading cell phone manufacturers were Sony, BlackBerry, Nokia, LG, Motorola, Samsung, and Sanyo. In addition, hundreds of independent phone companies, cable television operators, and Internet service providers offered local, long-distance, cellular, and voice-over-Internet phone services, as well as broadband Internet and cable television access, either singly or in bundled packages.

After several rounds of restructuring, AT&T emerged in 2000 as three distinct companies. One company focused on wireless (cellular phone service); the second focused on broadband (bundled Internet, phone, and cable television services provided via digital subscriber lines, known as DSL, which send digital data and voice transmissions over local telephone lines); and the third was AT&T, the surviving, traditional long-distance and local phone company that served businesses and consumers. AT&T was allowed to reenter the local phone business following the passage of the Telecommunications Act of 1996, which sought to increase competition among local, long-distance, and cable television providers. The new law allowed companies to get involved in all of these areas and offer bundled services.

AT&T had limped toward the twenty-first century battered and bewildered, but it emerged, in drastically altered form, reborn and in its customary position—the forefront of telecommunications. In 2005, it agreed to be purchased by SBC Communications (the former Southwestern Bell) for $16 billion. Shockingly, AT&T was no longer an independent corporation; however, out of respect for its long, dignified tradition, SBC changed its company name to AT&T, Inc. Although AT&T no longer had an exclusive hold on the telecommunications industry and was forced to compete with a large number of rivals, its enduring spirit of innovation, its institutional know-how, and its feeling of ownership over the communications network it had built allowed it to become an industry leader again. In some cases, it even earned goodwill and trust among customers who once resented its power. AT&T also benefited from consumers' familiarity with its name and respect for its long history of service, especially in a post-monopoly era of often unreliable and quickly bankrupted telecommunications service providers.

In 2007, this advantage allowed AT&T to become the exclusive carrier and distributor of the iPhone, Apple Inc.'s all-in-one, Internet-enabled, multimedia wireless mobile phone, which includes touch screens, an MP3 player, a simulated keyboard, a digital camera, e-mail, Web browsing, visual voice mail, text messaging, and local Wi-Fi connectivity. AT&T is still the largest communications company in the world. It offers cellular service to 70 million customers, and it remains the leading provider of long-distance and local wire-based phone service. It is the nation's largest provider of broadband, which includes Internet, telephone, and cable television services delivered via DSL.

ONE NATION, INDIVISIBLE AND FOREVER CONNECTED

Regardless of who the main telecommunications players turn out to be in the future and how the industry evolves, the infrastructure and systems that were developed following Alexander Graham Bell's first successful telephone call are an astounding American achievement. They are every bit as stunning and monumental as the Empire State Building, the Hoover Dam, the Interstate Highway System, and the Golden Gate Bridge, and of far greater, more practical value to billions of people worldwide, every day of their lives. Bell's pioneering work on the telephone led directly to the radio, television, satellites, cable television, cell phones, computers, and Internet technology. In short, Bell's transmitted words, "Mr. Watson! Come here. I want you," led directly to our twenty-first-century cyber culture of mass digital communication. The Information Age is built on Bell and Watson's humble transmitter and receiver.

The telephone first introduced Americans to the concept of communication without delay. Voices, gossip, news, information, business deals, and money could be sent and received in a split second. Results could be had right away. Space collapsed as time was condensed. What did distance matter if all personal and professional business could be conducted in real time, coast-to-coast and around the world? The bridging of space and the shortening of time quickly altered Americans' inner rhythms. They, too, sped up. As John Brooks describes it:

> In city and country alike, the telephone was creating a new habit
> of mind—a habit of tenseness and alertness, of demanding and

expecting immediate results, whether in business, love, or other forms of social intercourse. The twentieth century decline in the art of letter writing, which can legitimately be laid at the telephone's door, is only one symptom of the change of mood. The fact seems to be that the United States is the telephone's natural home and the twentieth its natural century, and that the instruments and the people found each other when the century was hardly begun.

A restless, impatient, results-oriented, driven people had found the tool that provided an outlet for their bottled energy and accurately echoed and amplified these national traits. If Brooks's theory is sound, today's cell phones would seem to represent an even more manic and hyperactive population, both connected to the world in unprecedented ways and increasingly disconnected from one another.

However, even in our current age of technological advances, Bell's greatest legacy remains the wiring of America—the drawing out of its most isolated and lonely citizens, and the gathering of the population into an intimate national conversation. A nation of remote farmhouses and homesteads, uprooted suburban families, and city dwellers separated by the anonymity of dense vertical living was suddenly connected.

Bell's invention, based on electrical current and sound waves, electrified America and got it talking. Although the more recent forms of instant communication that have arisen from telephone technology are new and different and always evolving, the energizing warmth of human give-and-take remains a constant. Like he had done for his deaf students, Bell unloosed America's tongue, and the resulting bonds forged between friends, family members, associates, and fellow citizens grew and remain strong.

CHRONOLOGY

1847   Alexander Graham Bell is born in Edinburgh, Scotland.

1876   Bell and his assistant, Thomas Watson, successfully test the world's first telephonic transmission. The prototype telephone is unveiled and successfully demonstrated at the International Centennial Exposition in Philadelphia, Pennsylvania.

1877   Bell receives a second patent for the telephone receiver, transmitter, and mechanical operations. Several hundred "point-to-point" phone lines begin to be set up in New England, primarily connecting businessmen's homes and

TIMELINE

1877
Bell marries Mabel Hubbard.

1847
Alexander Graham Bell is born in Edinburgh, Scotland.

1847 — 1878

1876
Bell and Thomas Watson successfully test the world's first telephonic transmission. The prototype telephone is unveiled and successfully demonstrated at the International Centennial Exposition in Philadelphia, Pennsylvania.

1878
Ten thousand Bell telephones are leased by phone subscribers.

offices. Bell, Watson, Hubbard, and Sanders form the Bell Telephone Company. Bell marries Mabel Hubbard.

1878   Ten thousand Bell telephones are leased by phone subscribers. Phone exchanges begin to appear, linking several dozen subscribers within small phone networks. Control of Bell Telephone shifts away from Hubbard and Sanders to Boston financiers. Theodore N. Vail becomes Bell Telephone's general manager.

1882   Bell Telephone acquires Western Electric, providing the phone company with a manufacturing arm for its telephone equipment.

1884   A New York-to-Boston phone line is established.

1977
The first fiber-optic cable to be used in a commercial communications system is laid.

1956
A transatlantic telephone cable is laid on the ocean floor and begins to provide high-quality phone service to Europe.

1996
The Telecommunications Act of 1996 spurs greater competition among local, long-distance, and cable television providers.

1915 ———— 2007

1915
The New York-to-San Francisco long-distance telephone line is opened by a test call between Watson and Bell.

2007
AT&T becomes the exclusive carrier, service provider, and distributor of Apple's iPhone.

1885 The American Telephone and Telegraph Company (AT&T) is created as a subsidiary of the American Bell Telephone Company.

1892 A New York-to-Chicago long-distance line is established.

1894 Bell's patent protection expires; an era of competition in the telephone industry begins. Within a few years, 6,000 independent phone companies are created.

1899 AT&T, based in New York City, acquires the assets of its parent company to take advantage of the more free-wheeling New York business environment.

1900 The loading coil is invented.

1915 The New York-to-San Francisco long-distance telephone line is opened by a test call between Watson and Bell.

1919 The first dial telephones are introduced, which allow customers to place calls directly without seeking operator assistance.

1925 Bell Labs, AT&T's research and development arm, is created.

1927 Transatlantic telephone service is begun using two-wave radio. Phone service to Paris follows in 1928, to Japan in 1934, and around the globe in 1935. AT&T stages the first public demonstration of television; two years later, it is the first to demonstrate color television.

1946 The first mobile phone, for use in cars, is introduced.

1947 Bell Labs invents the transistor.

1951 Customer-dialed long-distance calls begin to be introduced.

1956 A transatlantic telephone cable is laid on the ocean floor and begins to provide high-quality phone service to Europe. A transpacific cable is laid in 1964.

1958 AT&T introduces the first commercial modem.

1960 NASA launches the passive communications satellite *Echo I*, developed by Bell Labs. Two years later, Bell Labs' active communications satellite *Telstar I* is placed into orbit.

1963   Touch-tone push-button phones are introduced.

1977   The first fiber-optic cable to be used in a commercial communications system is laid.

1983   AT&T establishes a fiber-optic telephone route between New York City and Washington, D.C. It establishes the first cellular phone system in Chicago, Illinois.

1984   AT&T is broken up following the settlement of a government antitrust lawsuit. It loses control of the local Bells, and local and long-distance service is opened up to greater competition.

1988   The first transatlantic fiber-optic cable is laid.

1996   The Telecommunications Act of 1996 spurs greater competition among local, long-distance, and cable television providers.

2007   AT&T becomes the exclusive carrier, service provider, and distributor of Apple's iPhone. It is the nation's largest provider of broadband, which includes Internet, telephone, and cable television services delivered via DSL.

2009   Apple's iPhone dominates the emerging "smartphone" market. In 2009, one-third of all smartphone users owned an iPhone.

GLOSSARY

acoustics   Relating to hearing and sound as it is heard.

apparatus   An instrument or tool needed for a specific use; a device or machine designed for a specific use.

broadband   In telecommunications, a signaling method that includes or handles a wide range of frequencies that may be divided into many separate channels.

cellular telephone   Generally referred to as a "cell phone"; a battery-powered wireless telephone that sends and receives messages via radio waves transmitting within and among distinct service areas or "cells."

circuit   A complete or partial path through which electricity flows; any wiring hooked up to this path, such as for radio, telephone, and television transmissions.

communications   The science of transmitting information, messages, and signals.

current   The flow of electric charge along a conductor between two points.

digital subscriber line (DSL)   Digital information sent across a high bandwidth channel; enables cable television transmission, high-speed Internet access, and voice and other data transmission through fiber-optic cables.

electromagnetism   Magnetism produced by an electrical current.

exchange   A central office or system that provides telephone communication in a community or in part of a city.

infrastructure   The basic structures and facilities that make civilized society possible, such as roads, bridges, electrical plants and wires, schools, and communication and transportation systems.

innovation   Something newly introduced; a new and improved product, device, machine, technique, process, way of doing something, or way of thinking.

intermittent   Stopping and starting again at intervals.

modem   A device that converts data into a transmittable form and transmits it from a computer via a route (for example, telephone lines) to data-processing equipment on the other end, where the transmission is translated back to data.

monopoly   Exclusive control of a service or commodity within a given market that allows for the elimination of competition and the freedom to set high prices.

obsolete   Outdated, outmoded; no longer useful or practical.

patent   A document that grants the bearer the right to produce, sell, or profit from an invention.

prototype   A working model; an original model.

subsidiary   A company controlled and partly or wholly owned by another, usually larger, company.

telecommunications   Communication via telegraph, telephone, radio, television, computer, etc.

telephony   The science of telephonic transmission, of transmitting sounds across distances; the making or operation of telephones.

transmission   The passage of sound waves through space between a transmitter and a receiver.

Boettinger, H.M. *The Telephone Book: Bell, Watson, Vail, and American Life, 1876–1976*. Croton-on-Hudson, NY: Riverwood, 1977.

"A Brief History." AT&T. Available online. http://www.corp.att.com/history/history1.html.

Brooks, John. *Telephone: The First Hundred Years*. New York: Harper & Row, 1976.

Casson, Herbert N. *The History of the Telephone* (reprint edition). Fairfield, Iowa: 1st World Library Literary Society, 2004.

Cauley, Leslie. *End of the Line: The Rise and Fall of AT&T*. New York: Free Press, 2005.

"Central Office." InetDaemon. Available online. http://www.inetdaemon.com/tutorials/telecom/pstn/central_office/index.shtml.

Coll, Steve. *The Deal of the Century: The Breakup of AT&T*. New York: Touchstone, 1986.

Crandall, Robert W. *Competition and Chaos: U. S. Telecommunications Since the 1996 Telecom Act*. Washington, D.C.: Brookings Institution Press, 2005.

"Customer Premise Equipment." InetDaemon. Available online. http://www.inetdaemon.com/tutorials/telecom/pstn/cpe/index.shtml.

Farley, Tom. "Tom Farley's Telephone History Series." Privateline.com. Available online. http://www.privateline.com/TelephoneHistory/History1.htm.

Fischer, Claude S. *America Calling: A Social History of the Telephone to 1940*. Berkeley, Calif.: University of California Press, 1992.

Gross, Grant. "AT&T: Mobile and Broadband Strong in Q4: iPhone Customers Add to Subscriber Base." *Computerworld*. Available online. http://www.computerworld.com/

action/article.do?command=viewArticleBasic&articleId=
9058740&source=rss_news50.

"History of the AT&T Network." AT&T. Available online. http://
www.corp.att.com/history/nethistory/.

"History of the Public Switched Telephone Network."
InetDaemon. Available online. http://www.inetdaemon.com/
tutorials/telecom/pstn/history.shtml.

"Inventing the Telephone." AT&T. Available online. http://www.
corp.att.com/history/inventing.html.

Messenger, James R. *The Death of the American Telephone &
Telegraph Company: How "Ma Bell" Died Giving Birth to
the Information Age: An Eyewitness Account*. Marietta,
Ga.: The Alexander Press, 2007.

"Milestones in AT&T History." AT&T. Available online. http://
www.corp.att.com/history/milestones.html.

"North American Numbering Plan." InetDaemon. Available
online. http://www.inetdaemon.com/tutorials/telecom/pstn/
nanp/index.shtml.

Shulman, Seth. *The Telephone Gambit: Chasing Alexander
Graham Bell's Secret*. New York: Norton, 2008.

Smith, George David. *The Anatomy of a Business Strategy:
Bell, Western Electric, and the Origins of the American
Telephone Industry*. Baltimore, Md.: The Johns Hopkins
University Press, 1985.

Sterling, Christopher H., Phyllis W. Bernt, and Martin B. H.
Weiss. *Shaping American Telecommunications: A His-
tory of Technology, Policy, and Economics*. Mahwah, N.J.:
Lawrence Erlbaum Associates, 2006.

"Types of Phone Companies." InetDaemon. Available online.
http://www.inetdaemon.com/tutorials/telecom/pstn/phone_
companies.shtml.

Alphin, Elaine Marie. *Telephones* (Household History). Minneapolis, Minn.: Carolrhoda Books, 2001.

Bankston, John. *Alexander Graham Bell and the Story of the Telephone*. Hockessin, Del.: Mitchell Lane, 2004.

Banting, Erinn. *Inventing the Telephone* (Breakthrough Inventions). New York: Crabtree, 2006.

Besing, Ray G. *Who Broke Up AT&T?: From Ma Bell to the Internet*. Bloomington, Ind.: 1st Books Library, 2000.

Dooner, Kate E. *Telephones: Antique to Modern*. Atglen, Pa.: Schiffer, 2004.

Evenson, A. Edward. *The Telephone Patent Conspiracy of 1876: The Elisha Gray–Alexander Bell Controversy and Its Many Players*. Jefferson, N.C.: McFarland & Co., 2001.

Fandel, Jennifer. *Alexander Graham Bell and the Telephone*. Mankato, Minn.: Capstone Press, 2007.

Kummer, Patricia K. *The Telephone* (Inventions That Shaped the World). New York: Franklin Watts, 2006.

Martin, Dick. *Tough Calls: AT&T and the Hard Lessons Learned from the Telecom Wars*. Saranac Lake, N.Y.: AMACOM, 2004.

Mattern, Joanne. *Telephones* (Transportation and Communication Series). Berkeley Heights, N.J.: Enslow, 2002.

McCormick, Anita Louise. *The Invention of the Telegraph and Telephone*. Berkeley Heights, N.J.: Enslow, 2004.

Meyer, Ralph O. *Old-Time Telephones!: Design, History, and Restoration*. Atglen, Pa.: Schiffer, 2005.

Stefoff, Rebecca. *The Telephone* (Great Inventions). New York: Benchmark Books, 2005.

Webster, Christine. *Alexander Graham Bell and the Telephone* (Cornerstones of Freedom). New York: Children's Press, 2004.

Williams, Brian. *Bell and the Science of the Telephone*. New York: Barron's Educational Series, 2006.

PICTURE CREDITS

INDEX

ABOUT THE AUTHOR

JOHN MURPHY is a writer and editor of young adult books. He holds an M.A. in medieval literature.